Echoes of the North-East Miners

Some last traces of the collieries
and tributes to the pitmen

Ken Smith and Dr Tom Yellowley

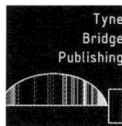

Tyne
Bridge
Publishing

Published by:
City of Newcastle Upon Tyne
Newcastle Libraries
Tyne Bridge Publishing, 2019
www.tynebridgepublishing.org.uk

Seaton Delaval spoil heaps:
A distant view of Seaton Delaval Colliery, Northumberland, showing its spoil heaps. The mine closed in 1960. Spoil, or slag, heaps were a prominent feature of the North-East landscape.

Introduction

When Ellington Colliery, Northumberland, closed in January 2005 the event ended around 800 years of continuous coal mining in the North-East of England, apart from a few opencast sites.

The pits of the Great Northern Coalfield had led the way in fueling the engines which made Britain a powerhouse of manufacturing and innovation during the Industrial Revolution, and those pits continued to make a fundamental contribution to Britain's economy well into the 20th Century.

As well as industry, the miners of Northumberland and County Durham supplied heating and lighting to millions of homes throughout Britain. They literally kept the home fires burning.

Black diamonds were also mined for the gas industry from Victorian times onwards and during the 20th Century became the bedrock of Britain's electricity power stations.

The United Kingdom was not the only market for the coal which the pitmen - and pit lads - of the North-East wrested from the deep earth by their arduous labour. It found its way to destinations worldwide.

This achievement came at a tremendous cost in terms of life and limb. The miners - many of them under the age of 20 - risked death or injury on a daily basis to win the coal from the dark seams far beneath the surface.

Methane gas - known to the pitmen as firedamp - could cause major explosions, resulting in great loss of life. The firedamp ignited when exposed to naked lights, such as candles or unprotected oil flames. Shot-firing operations to bring the coal down from the face might also sometimes ignite this gas.

Coal dust could greatly increase the power of such blasts and could itself be explosive, although this was not fully understood in earlier years.

Deadly carbon monoxide gas - known as afterdamp - often formed in the wake of firedamp explosions, adding to the death tolls. Chokedamp (very low oxygen and high carbon dioxide levels plus nitrogen) might also be present in the aftermath, and this too was a serious peril. It produced a suffocating atmosphere.

Chokedamp, also known as blackdamp, did not just occur after explosions. It was also generated by animals and people breathing, the burning of lamps and candles and the slow oxidation of coal – the latter sometimes occurring as 'outbursts'.

Other dangers included falls of stone which would frequently kill or injure a man or boy. In addition, transport accidents involving coal tubs were not uncommon, although falls of stone were more numerous.

Water was another hazard. It could - at great pressure - burst in on miners from old, flooded workings. Examples of such flooding include the disasters at Heaton Main Colliery in 1815 (75 dead) and at Montagu View Colliery in 1925 (38). Both these pits were in the Newcastle area. A memorial to the Montagu View dead can be found in the city's St John's Cemetery, Elswick, overlooking the River Tyne.

Durham Mining Museum researchers have so far discovered around 20,900 known on-site deaths at collieries in County Durham and Northumberland, although this number excludes those countless miners who died from mining-related diseases such as pneumoconiosis and emphysema and those who died days or months after receiving injuries.

In addition, before 1850 there was no legal requirement to record fatal colliery accidents. The numbers of men and boys who lost their lives must therefore be very much greater than 20,900.

However, safety measures in the pits were gradually improved, particularly during the 20th Century. Greatly enhanced ventilation using powerful fans to expel gas and the introduction of hydraulic steel pit props were two factors which brought down the death tolls. The pitmen's unions continually pressed for better safety measures.

Despite such improvements, mining remained a dangerous occupation. As late as 1951, 81 miners and two rescue workers died as the result of the Easington Colliery disaster, which was caused by a devastating explosion.

Two of the last men to be killed in the Durham coalfield

The relief carving of a miner on the main memorial to the 81 men and boys who died in the Easington Colliery disaster of 1951. Two rescue workers also lost their lives.

lost their lives as the result of a transport accident at Wearmouth Colliery in February 1992.

Today, there are few signs of the vast number of collieries which existed in County Durham, Tyne and Wear and Northumberland. The overwhelming majority have been erased from the landscape. A considerable number of their sites are now occupied by developments such as country parks, housing and business estates.

Only a handful of pithead winding wheels - or pulley wheels as they are often called - survive in their original position. Yet during the 19th and early 20th centuries they dotted the landscape in profusion.

Similarly, nearly all the colliery spoil heaps have been removed, smoothed away by bulldozers or transformed into green hills by landscaping. These mounds of shale, small particles of coal, stones and dust glowed at night like volcanoes and gave off smoke as the result of internal combustion.

Occasionally, capped shafts can be seen - often visible as areas of concrete - and sometimes pipe-like vents to allow for the escape of mine gas are in evidence at these sites. But not all old shafts have been concreted over - a number are covered with metal grills. Air shafts are also detectable at various locations.

However, hidden from our eyes are the countless miles of underground mine workings which lie beneath the streets and fields of the North-East. Many are flooded.

On the surface, a considerable number of former tied pit housing rows, built by the colliery companies, survive. Examples include those near the Stadium of Light in Sunderland (the site of Wearmouth Colliery), at Chopwell near Gateshead, at Brunswick Village (site of Dinnington Colliery) near Wideopen, at Boldon Colliery near South Shields, at Ashington in Northumberland and at Blucher Village on the western side of Newcastle. There are many other examples of these terraces, particularly in County Durham.

Today, of course, almost all the homes have been modernised.

Also found in communities throughout much of the North-East are numerous memorials to the miners, their collieries and to the pitmen and lads who lost their lives in the many disasters. Pit winding wheels and coal tubs are among the most numerous symbols adopted as memorials. Tubs and wheels are often positioned on roadside grass verges or in other prominent positions in the former pit villages and towns, sometimes at the site of the vanished colliery.

A number of impressive memorial sculptures have been erected, including those at Ellington in Northumberland (statue of miner), at Concord, Washington (statue of miner, his wife and son), at Esh Winning (statue of miner, his wife and small daughter) and Horden (statue of miner).

Disaster memorials include New Hartley (204 deaths), West Stanley (168), Seaham (164), Wallsend (marked by plaque, 102), Haswell (95), Felling (92), Easington Colliery (83), Trimdon Grange (74), St Hilda, South Shields (51) Montagu View, Newcastle (38) and Stargate (38).

Thousands of former miners and their descendants still live in County Durham, south east Northumberland and Tyne and Wear. The large number of memorials shows that they, and their families and friends, have ensured that the proud heritage of the region's pitmen is not forgotten. Indeed, the abundance of such tributes is a remarkable feature of the North East.

These memorials, together with the few traces of the Great Northern Coalfield which we can see today, are a reminder of an industry which shaped the character of the North-East and its people like no other.

Aged Miners' Homes

In addition to the memorials, the North East's towns and villages often feature another visible sign of how important the miners were to the area. These are the rows of single-storey Aged Miners' Homes which are to be found in many former pit communities throughout much of County Durham, Tyne and Wear and south east Northumberland. Indeed, they are such a familiar part of the North-East scene that they are frequently overlooked. These homes frequently carry inscriptions such as "Haven of Rest" and words commemorating their opening.

The Aged Miners' Homes movement proved to be a complete success story, showing the charity and concern of the miners towards their fellow pitmen. Other benefactors, including in some cases the coal owners themselves, contributed money, land and materials, but the miners were the main source of regular funding.

The movement to provide homes for retired miners and their wives sprang largely from the fact that many pitmen lived in housing owned by the colliery companies.

These tied tenancies meant that once a man ceased to work at the colliery he and his wife were liable to lose their home, unless they had sons still living with them and working at the mine. It was likely that the couple would be unable to afford decent alternative accommodation. In addition, those miners living in non-colliery houses would be unable to afford the rents when they retired.

Adair Terrace at Chopwell:
This long row of retirement bungalows, siutated on a hillside at Chopwell, was completed by the Durham Aged Mineworkers' Homes Association in 1923. It was named in honour of John Adair, a secretary of the association for many years. The homes bear stone plaques on their facades inscribed with the names of people and organisations which helped with the funding of the development.

Examples include James Gilliland, representing Chopwell's Miners' Union Lodge and Harry Bolton, secretary of the lodge, representing Blaydon Co-operative Society. Other names include representatives of the Methodists, the villages's Engineeers' union lodge, Hamsterley miners' lodge, Chopwell institute, Hamsterley Club and colliery officials. Indiviuals include a colliery manager, a Church of England clergyman and Willie Whiteley, Labour MP for Blaydon.

The movement began in the Durham Coalfield. A trail-blazing initiative was taken by miners of the union lodge at Boldon Colliery who in c.1896-98 acquired and opened Down Hill House at West Boldon, a large, disused property which had been converted to become accommodation for retired pitmen. This set an example for others to follow.

The Durham Aged Mineworkers' Homes Association (DAMHA) was formed in 1898 under the leadership of Joseph Hopper, a Gateshead pitman and Primitive Methodist lay preacher who became its first secretary. He was supported by miners' MP and Durham Miners' Association (DMA) general secretary John Wilson, the DMA financial secretary John Johnson, the Bishop of Durham Brooke Foss Westcott, Canon William Moore Ede and Henry Wallace.

Today, the DAMHA provides around 1,700 homes, mainly for retired people. The association points out that the foundation of this organisation was the result of the vision of Joseph Hopper, who believed that a man who had served in the coal mines from the age of 12 to 65 or beyond deserved better than to be evicted from his tied colliery house when he retired without the prospect of a decent alternative home. Such accommodation was vitally needed for elderly pitmen and their wives.

The earliest homes provided by the DAMHA were existing properties no longer required by working miners. The first group to be opened were at Haswell Moor, in 1899, where the association had purchased the complete village for housing following closure of the pit. Similar schemes involving existing houses followed at Shincliffe, Shotton Colliery and Houghall.

The initiative gained momentum. In the early 1900s, houses were built in almost every union lodge area in County Durham and according to the association by the outbreak of the First World War in 1914 a total of 475 homes had been constructed and a number of single men's hostels provided.

Among the first purpose-built properties were 12 cottages completed in the Wrekenton-Springwell area in 1904. The little settlement was known as "Wallace Village", being named after Henry Wallace, one of the movement's pioneers.

The Durham miners' union lodges were key backers of the housing movement, the men contributing money from their pay on a regular basis. The finance provided by the miners was fundamental to the success of the association's efforts. In the early years of the movement the homes were rent-free. This was in no small measure due to the regular donations of the pitmen.

The miners were not, however, the only source of finance and help. Organisations of many kinds, including some colliery companies, helped to provide the homes, with funding, and sometimes land and materials. A collection for the DAMHA was and still is made at the annual Miners' Festival Service in Durham Cathedral, held on the afternoon of the Durham Miners' Gala.

The 1920s saw many homes built, with between 50 and 100 completed annually. The DAMHA declares: "Although the Durham coalfield is no more, the association has survived and prospered, and continues to provide good quality homes for older people, the less physically active or disabled." It is no longer necessary to have been a mineworker.

The organisation is the largest North-East-based retirement homes provider. The organisation now concentrates on providing as many two-bedroom cottages as possible. The early homes were generally one-bedroomed.

However, from the 1960s onwards the closure of mines led to a decrease in the funding from pitmen's union lodges which had been such a vital source of income. In 1982, the DAMHA became a registered housing association with the Housing Corporation and this gave it access to government grants for new building and modernisation. But it retained its charity status. It is the largest Almshouse Charity in Britain.

The movement born in County Durham spread northwards. In 1900, an association was formed to provide homes for retired Northumberland miners and their wives. The Northumberland Aged Mineworkers' Homes Association opened its first row of single-storey cottages at East Chevington in 1902, a village which eventually disappeared with the closure of its drift mine.

As in County Durham, members of the Northumberland miners' union paid a levy from their wages to help fund provision and maintenance of the homes. Today, the organisation provides around 540 homes for former miners, their wives or widows.

The terraced bungalows at East Chevington were followed by similar homes in many other areas of Northumberland and Newcastle.

A group of 10 bungalows - or "cottages" as they were sometimes described - opened in the Klondyke area of Cramlington in 1928. They were dubbed the "Footballers' Group". Money to help build the homes had been raised by sponsored football matches. They were opened by the secretary of the Football Association and one of the foundation stones was inscribed with the name of Hughie Gallacher, the legendary Newcastle United player.

In Northumberland, the tenants of the earliest bungalows were chosen by drawing lots. This was seen as the fairest

Footballers' Group:
Some of the group of ten bungalows or cottages as they are sometimes described - opened in the Klondyke area of Cramlington in 1928. They were dubbed the 'Footballers' Group'. Money to help build the homes had been raised by sponsored football matches. They were opened by the Secretary of the Football Asscociation, and one of the foundation stones was inscribed with the name of Hughie Gallacher, the legendary Newcastle United football player.

method of allocating the cottages and was akin to the traditional drawing of lots known as cavilling, which decided a miners' workplace in the pit.

In both County Durham and Northumberland groups of homes were sometimes built in memory of people who had given valued service to the Aged Miners' Homes movement or were held in high regard by the pitmen.

Examples include the Burt Homes at Choppington, named after Thomas Burt, the leader of the Northumberland miners' union during the 19th Century, the Joseph Hopper Memorial Homes at Windy Nook and Joseph Hopper Homes at Birtley, the Adair Terrace homes at Chopwell, named after John Adair, a long-serving secretary of the DAMHA, and the Thomas Henry Cann Memorial Homes in Blue House Lane, Washington. Cann was a leading Durham miners' union official.

New Hartley Memorial Cottages:
The New Hartley Memorial Cottages, completed by the Northumberland Aged Mineworkers' Homes Association in 1910 at New Hartley. They commemorate the 204 men and boys who died in the North East's worst pit disaster in 1862.

Joicey Aged Miners' Homes, Herrington Burn:
The Joicey Aged Miners' Homes at Herrington Burn, near Shiney Row, opened in 1906 by the DAMHA. Attractive details makes this row of cottages particularly eye-catching. They are Grade II listed.

Joseph Hopper Aged Miners' Homes, Birtley:

The Joseph Hopper Aged Miners' Homes at Birtley, near Gateshead. These homes were built by the Durham Aged Mineworkers Homes Association (DAMHA) and commemorate the man who played a key role in founding the association and ensuring its success. Joseph Hopper was a miner and Primitive Methodist lay preacher from Windy Nook, Gateshead, who became the organisation's first secretary.

A second group of homes were also built in remembrance of this housing pioneer. They are the Joseph Hopper Memorial Homes at Windy Nook.

Joseph Hopper, who died in 1909, is buried in St Alban's Churchyard, Windy Nook, his grave surmounted by a memorial column, erected by the DAMHA in "affectionate and grateful remembrance". The inscription on the column also pays tribute to "his rare gifts of mind and heart" and his "untiring and self-denying labours".

Neat bungalows, Deaf Hill:

A neat row of miners' retirement bungalows at Deaf Hill, County Durham. Built by the DAMHA, this group of 10 properties was opened in 1926-27. Deaf Hill Colliery closed in 1967.

Aged Miners' Homes, Metal Bridge, Tursdale:

A group of homes built by the DAMHA at Metal Bridge, in the Tursdale area. They are dedicated as a memorial to the miners of collieries in the surrounding district who lost their lives in the First World War. These bungalows were completed in 1921.

In sight of pulley wheels:
The Betty Priestman Cottages, built by the Northumberland Aged Mineworkers' Homes Association close to the site of Woodhorn Colliery on the outskirts of Ashington. These 10 bungalows, typical of the many Aged Miners' Homes in the south east of the county, were opened in 1912. They have been fully updated and renovated. The preserved and restored winding wheel towers of the colliery, now part of a fine museum, are only a short distance to the rear of the homes.

Aged Miners' Homes pioneer:
The gravestone of Joseph Hopper, pioneer of the Durham Aged Miners' Homes movement, in St Alban's Churchyard, Windy Nook, Gateshead.

Funded by the pitmen:
(Centre) Aged Miners' Homes at Kibblesworth, near Gateshead. They are part of a row built by the DAMHA between the two world wars. (Left) On the rear walls of these bungalows are stone plaques commemorating miners' union lodges which helped to fund the development. They included Kibblesworth, Washington 'F', Washington Glebe and

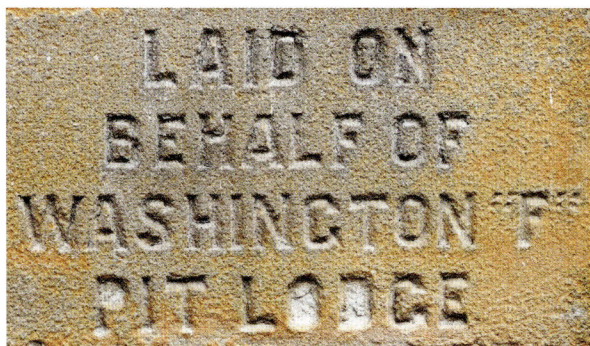

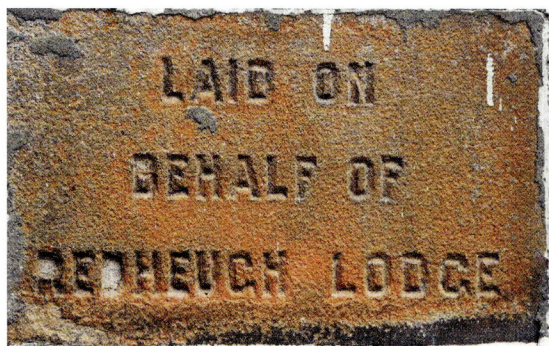

LAID ON BEHALF OF WASHINGTON 'F' PIT LODGE

LAID ON BEHALF OF REDHEUGH LODGE

Buildings, winding wheels, tunnels and waggonways

Winding engine house, Scremerston:
At Scremerston, near Berwick, the winding and pumping engine houses of the Victorian Scremerston Old Colliery survive in woodland on the southern edge of Scremerston village. They date to 1840. Pictured is the winding engine house, which is a very rare survival.

Pumping engine house, Scremerston:
The Scremerston Old Colliery pumping engine house. The engine houses are only a few yards from one another. A capped shaft-top is also visible. A later mine, known simply as Scremerston Colliery, was sunk on the northern outskirts of the village and survived until the 1940s. The various pits in this area were among the most northerly in Northumberland. A stone plaque on the side of one of the engine houses at Scremerston records the date they were built - 1840. The pit was owned by the Greenwich Hospital Commissioners.

Woodhorn winding wheels:

Above are pictured the beautifully preserved winding wheel towers of Woodhorn Colliery on the outskirts of Ashington, Northumberland. The Woodhorn pit is now the site of a mining museum and archives centre. The colliery closed in 1981, but the winding wheels and buildings have been retained as a major feature of the museum.

Among the buildings are the two winding engine houses. The No. 2 house contained one of the last steam-operated winding engines in Britain. However, in 1975 it was replaced by an electrically-operated engine from the Fenwick Pit at Earsdon in North Tyneside. This electric engine is still in place in the No. 2 house and can be seen in operation on some days.

A winding wheel from the Duke Pit of Ashington Colliery is also on display near the museum's main entrance.

Other pictures show the Woodhorn Colliery pithead around 1960, a winding drum in the No. 2 engine house, and the memorial statue to the 13 men who lost their lives at Woodhorn Colliery in 1916 as the result of a firedamp explosion. In recent years the memorial to these tragic miners has been moved from the car park area of the museum to a position closer to the winding wheel towers.

Clockwise from top left:
1) The statue of a miner on the memorial to the 13 miners who died in the Woodhorn Colliery disaster of 1916.
2) A winding drum in the number two engine house at Woodhorn.
3) The Woodhorn Colliery pithead pictured around 1960. The mine closed in 1981.

Beamish winding wheels and engine house:

South of the Tyne, the North of England Open Air Museum at Beamish features a pair of pulley wheels, an engine house and an 1855 winding engine, although they are not in their original position. The wheels and engine house were transferred to the site from the Beamish 2nd Pit, which was situated less than a mile away.

Other attractions at the museum include the Mahogany Drift mine, which dates back to the 1850s. A small section of the workings is open to visitors on guided tours. There is also a reconstructed miners' lamp cabin.

A memorial tablet to the 20 pitmen and boys who died as the result of an explosion at Brancepeth Colliery in 1886 can be found in the reconstructed band hall at the museum's colliery village. The youngest victim of the disaster was aged 14. The tablet was originally located at Willington, County Durham.

The pithead at Beamish Museum.

The entrance to the Mahogony Drift mine at the museum.

Washington F winding wheels:

A pair of winding or - as they are often called - pulley wheels and the engine house at the site of Washington F Pit. The Victorian winding engine, built in 1888, is still in working order, but is now operated by electric instead of the original steam power. Washington F Pit, in the Durham Coalfield, was one of the North-East's oldest collieries, dating back to the 18th Century. It closed in 1968.

Ushaw Moor winding engine house:

Below right: a small winding engine house at the site of Ushaw Moor Colliery in County Durham. This featured an electrically-operated winder. The shaft was only around 370ft deep, making it a relatively shallow mine. This winding engine house was used for pumping operations following the closure of Ushaw Moor Colliery in 1960. The mine dated back to 1865. The seams worked were the Harvey, the Yard, the Busty and the Brockwell.

Today, the winding engine house stands in a small wood of mainly fir trees, and is situated a short distance to the west of the village.

Ford Moss chimney:

The chimney of an engine house at the site of Ford Moss Colliery in North Northumberland. Mining in this area dated back to the 17th Century. Ford Moss is now a nature reserve, but the brick and stone chimney and other remains of its coal mining past have survived. Ford Moss Colliery closed in c.1918.

Blucher Colliery village:

Pictured top are colliery housing terraces which survive in Blucher Village on the western side of Newcastle. The Blucher pit was sunk in 1800 and closed in the 1860s. However, it was reopened in 1901.

The rows of homes for the miners and their families, were built in the early 1900s by the Throckley Coal Company. The houses have been updated. They are Stephenson, Spencer, Simpson, Boyd and Blucher terraces. The colliery manager's house was in Blucher Terrace, which stands close to the site of the pithead.

Two other terraces, Coquet Buildings, which featured two shops, and West Spencer Terrace, were not owned by the coal company. They have also survived. The village still has a social club, which is situated across the main road from the homes. The present building opened in 1931 but the club dates to 1919.

The two other photographs show Blucher's fine Methodist chapel, the gift of Sir William Stephenson, head of the coal company. The chapel was opened close to the terraces in 1906 and is still in use. It contains an excellent organ and two stained glass window panes rescued from the now demolished Throckley Methodist Chapel.

A narrow-track tubway operated by a stationary engine took the coal from the Blucher mine down to Lemington on the banks of the Tyne. A locomotive-operated railway from North Walbottle Colliery ran alongside part of the tubway and also terminated at Lemington.

Final closure of Blucher Colliery came in early 1956. The mine and village were named in honour of the Prussian General Blucher of Waterloo fame. The village was once visited by a party of German tourists who were curious as to why it should bear the name of the famed soldier who played a prominent role in the later stages of the Battle of Waterloo.

Pit housing rows, Chopwell:

Former colliery housing rows at Chopwell. The streets are all named after British rivers, including the Tyne, Tweed and Humber. Chopwell Colliery opened in 1896 and closed in 1966. During the Great Lockout and General Strike of 1926, when Britain's miners were battling against proposals for longer working hours and reduced pay, Chopwell pitmen gained a reputation for their militant opposition to the plan. This led to the village, somewhat unfairly, being dubbed "Little Moscow".

Pit housing rows, Brunswick Village:

Former pit housing rows at Brunswick Village, the site of Dinnington Colliery. The village itself was at one time known as Dinnington Colliery. The mine was situated in what is now a small industrial and business estate, across the road from the housing rows.

Dinnington Colliery closed in 1960 after more than 90 years of operation. Other nearby pits included Havannah, Hazlerigg, Seaton Burn and Brenkley. Dinnington village, about two miles from Dinnington Colliery, had its own mine, known, rather confusingly, as East Walbottle Colliery.

Deputy Row, Scremerston:

Deputy Row at Scremerston, near Berwick. This group of terraced homes was built, as its name implies, for deputy overmen and their families. Deputies were in charge of safety in districts of a mine, leading the teams of men working at the coalfaces.

Langley Park Colliery workshops:

A colliery workshops building (pictured top right) has survived at the site of Langley Park Colliery, now a small industrial estate. The workshops were used by the mine's electricians and blacksmiths. Some of the pit's rail lines can be seen outside the building. Nearby the brick-built pithead baths survived until quite recently, although they were due to be demolished.

Langley Park Colliery, in the Durham Coalfield, opened in 1873 and closed in 1975. The mine went down to around 450ft.

Also in the Durham Coalfield, former mine buildings which survive include those at the sites of Sacriston, Craghead, Kimblesworth, Wardley and Lumley Sixth Pit collieries. At the location of the former entrance to Harton Colliery in South Shields buildings once used as the manager's house, under-manager's house, overmen's houses and pay and time-ticket office still stand in Loudon Street, a short road next to the site of the colliery entrance. In the Northumberland Coalfield, pithead buildings still in existence include those of Woodhorn Colliery, on the outskirts of Ashington, and Fenwick Colliery, near Earsdon.

Yet all these and the other buildings represent only a fraction of the mines which have existed in the region over the centuries. The records of the North East are littered with the names of forgotten or half-forgotten pits.

Brenkley Colliery workshops:

The well-preserved workshops building of Brenkley Colliery on the northern outskirts of Newcastle. The workshops, which are today used as offices by various organisations, are alongside the site of Seaton Burn Colliery, now the site of an offices and business estate. The Brenkley drift mine was an extension of the old Seaton Burn pit workings and was situated only about a mile to the west of Seaton Burn Colliery.

Brenkley rail track way:

The old track way for the wire-rope haulage railway which took the coal from Brenkley Drift to Seaton Burn Colliery. It was then transported onwards through the mineral railway system to staiths on the Tyne. The seams worked were the Bensham and the Low Main. Brenkley closed in 1986.

Coxlodge waggonway:

In 1605-1608 the earliest known horse-operated wooden waggonway to transport coal in the North East was opened in the Blyth area. Waggonways were to supersede horse-drawn carts and packhorses which were impeded by the appalling state of roads and tracks in earlier times. By the middle of the 18th Century a network of these waggonways laid with wooden rails had developed to take the coal from the pits to the staiths on the rivers Tyne, Wear and Blyth.

Pictured is part of the surviving line of the Coxlodge Waggonway in Newcastle. Originally known as the Kenton and Coxlodge Waggonway, at various periods it served the East Kenton, Coxlodge and Gosforth collieries and ran down to staiths at Wallsend.

Today, part of the line of the track survives as a wide path from near Benton Park Road, close to Haddricksmill Bridge at South Gosforth, to Benton Road. On the other side of Benton Road it continues to Coach Lane and onwards to Meridian Way. A section of the route passes between the Freeman Hospital and the Government buildings of the Department of Work and Pensions. In its early years the waggonway was operated by horses.

In 1812, Tyneside-born colliery viewer (manager) John Blenkinsop and engineer Matthew Murray completed a "rack" locomotive for a mine railway in the Leeds area. Murray was probably its designer with Blenkinsop patenting the rack-rail system. The driving wheel featured cogs which slotted into a rack rail as the engine moved along, but its action was slow and lop-sided.

This awkwardness of motion did not, however, prevent the engine from being tested on the Kenton and Coxlodge Waggonway. At least two of these Blenkinsop-Murray locomotives were tried out for a short time on the route in c.1813. In later years, the line featured conventional rails and locomotives. Fixed engines operating wire-rope haulage were also used on this route.

Kenton Colliery capped shaft:

A capped main shaft of Kenton Colliery, sometimes known a East Kenton Colliery, can be seen at the southern end of Moor Lane, Kenton, Newcastle. It is located next to a group of trees close to Bowfell Avenue and Bedale Green. This concrete area marking the shaft is situated on top of a mound characteristic of early pit sites. It is the only visible sign of the colliery, which closed during the first half of the 19th Century.

Mining at Kenton dated back to the 16th and 17th centuries. There is a reference to coal pits in the district as early as 1577 and documents from the 1680s mention Kenton Colliery.

A three-mile-long tunnel, known as Kitty's Drift, ran from the colliery to staiths on the Tyne near Scotswood. Completed in 1796, this underground passage had a dual role - it was a drain to remove water from the pit and also carried a pony-operated railway for coal waggons. The entry to Kitty's Drift may have been situated near the foot of the now capped shaft.

The workings of Kenton Colliery were linked underground to those of the nearby Coxlodge Colliery, which opened in 1810. After the run-down of the Kenton coal reserves, the Kenton pumping engine was used to help drain water from the Coxlodge workings. Today, it is difficult to visualise that Kenton, a suburb of Newcastle, was formerly a pit village occupied mainly by miners and their families.

Wardley pit ruins:

A ruined building at the site of Wardley Colliery, No 2 and No 3 shafts, at White Mare Pool on the eastern fringes of Gateshead.

Wardley Colliery opened around 1800-1802 and was worked until 1911 when it was closed and replaced by the nearby Follonsby Colliery, a little to the south. In 1958-59, however, Wardley Colliery was reopened and merged with the Follonsby mine. The original Wardley Colliery shafts were from then on known as Wardley No. 2 and No. 3. The Follonsby shaft was named Wardley No. 1 and linked to the Usworth mine by an underground locomotive road. All mining at the Wardley-Usworth combine ceased in 1974. At the time of writing a number of other ruined buildings survive at the Wardley site, including the offices and canteen. The photo was taken in 2013. However, a housing development is now planned for the site.

Lynemouth pithead baths:

The pithead baths at the site of Lynemouth Colliery in Northumberland. This Grade II listed building was completed in the 1930s. The small tower helped to provide the head of water needed for the showers. Lynemouth Colliery closed in 1994. The seams, which extended under the North Sea, included the Diamond, High Main, the Main, the Yard and the Brass Thill. The pit was only a short distance from Ellington Colliery.

Lynemouth shower cubicles:

A view of the tiled shower cubicles for the miners inside the Lynemouth pithead baths. Although there was a report that the owners of Cramlington Colliery had provided miners with baths at the pithead as early as 1855, this was an exceptional case. Not until the 20th Century did a determined movement get underway towards providing these facilities.

The first "modern" pithead baths in Northumberland were opened at Ellington Colliery in 1924 and the first in County Durham at Boldon Colliery in 1927.

Lynemouth lockers:

The still-intact lockers at the Lynemouth baths. It took many years for most of the North-East's mines to be equipped with baths. These shower facilities were a great leap forward for the miner's quality of life and that of his wife. The pitman was no longer forced to return home covered in black grime and dust and his wife was saved the daily chore of filling and emptying a tin bath.

Cambois pithead baths:

The Cambois Colliery pithead baths on the Northumberland coast near Blyth. A memorial to the colliery and its miners takes the form of a pithead winding wheel and tub. The pithead baths can be seen behind the tub. The Cambois mine opened in the 1860s and closed in 1968. Its workings ran under the North Sea.

Immediately to the south of Cambois is the port of Blyth, from where coal was exported in large quantities. Blyth was very much a mining town, with pits such as Crofton Mill (generally known simply as 'Mill'), Isabella and New Delaval. One of the earliest mines in the town was Cowpen and this was succeeded by the nearby Bates Colliery. Bates became the last mine in Blyth, closing in 1986.

Elemore pithead baths:

A view of the impressive-looking pithead baths at the site of Elemore Colliery in the Durham Coalfield. As with all such baths, the water tower is a distinguishing feature. Elemore Colliery, located near Easington Lane, closed in 1974 after more than 100 years of mining. This large building, which dates to the 1930s like those at Lynemouth, retains original tiling and traces of the partitions separating the shower cubicles. Unusually, the showers were arranged on two floors.

Elemore Colliery was situated near Easington Lane to the south of Hetton-le-Hole. Its site is now occupied by a golf course.

Other surviving pithead bath buildings include those at Fishburn in County Durham, where the medical block can also be seen; the site of the Fenwick Colliery, near Earsdon and Backworth; and Weetslade, near Wideopen.

Addison tunnel:

Top right, the entrance to the rail tunnel close to the site of Addison Colliery, which was situated next to the Tyne, not far from Ryton. The railway ran through the tunnel under the Blaydon-Ryton road and linked up with Stargate Pit at Ryton, Emma Colliery on the edge of Crawcrook and, for a time, Greenside Colliery. The railway took the coal from these mines down to the main line and onwards to the staiths. Other collieries within a few miles of Addison included Clara Vale, Catherine, Blaydon Mary and Blaydon Bessie.

Victoria Tunnel:

A photograph taken in the Victoria Tunnel, which runs under part of Newcastle city centre. The tunnel was built to carry coal from the Leazes Main Colliery at Spital Tongues to a point near the mouth of the Ouseburn, close to the staiths on the Tyne.

The Victoria Tunnel was constructed in 1839-1842 and runs beneath Claremont Road, the Great North Museum site, Newcastle Civic Centre and the Shieldfield district to a point below Byker, close to where the Ouseburn meets the Tyne.

The full waggons used a descending gradient to reach the riverside. The empty waggons were then hauled back up the tunnel by a cable-winding engine at the pit. This underground passage has a maximum depth of 85ft and was used as an air raid shelter during the Second World War. A number of extra entrances were created for this purpose.

Coal mining also took place for hundreds of years on Newcastle's Town Moor. Coal was opencasted there as late as 1944-1947. Today, ring-shaped banks of spoil survive on the Moor from the early bell-pits phase of mining. The tops of pit shafts are discernible on the Town Moor, Nuns Moor and Little Moor. In the 17th Century, the 'Town Moor Colliery' was said to have extended for 100 acres under the grassland.

St Hilda's Colliery pithead building:

Pictured below is the building housing the shaft at the site of the St Hilda's Colliery in South Shields. This pithead building survives as a memorial to the miners and their colliery.

Coal-drawing ceased at St Hilda's in 1940. The shaft was afterwards used for ventilation and as an emergency exit for Westoe Colliery. Westoe, also at South Shields, closed in 1993 and was the last working pit on Tyneside.

The surviving pithead shaft building of St Hilda's has undergone restoration work led by the Tyne and Wear Building Preservation Trust. This has created three office/studio spaces from the former lamp room, two meeting/exhibition/event rooms and other spaces for exhibitions and events. The area of the capped shaft around the pit cage is to be used for an exhibition of mining heritage items. The pulley wheels near the top of the building have also been restored.

The St Hilda's Colliery disaster memorial, which takes the form of a pithead wheel, can be seen nearby next to the Asda store in the town. Fifty one men and boys died in the tragedy in 1839. It was the result of a devastating explosion. The disaster led to the setting up of a South Shields committee to investigate the causes of accidents in coal mines and to propose improvements.

The committee recommended that all mines should have at least two shafts and stressed that poor ventilation was a major hazard. St Hilda's dated back to the early nineteenth century.

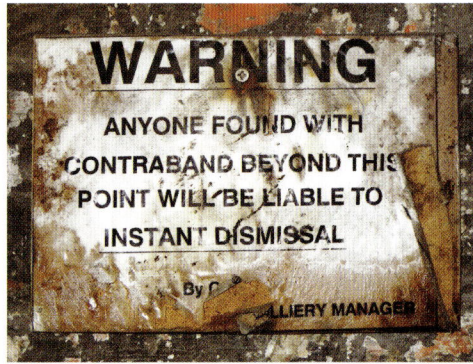

St Hilda's Colliery cage:

Inside the shaft building at the St Hilda's site is the two-deck pit cage (pictured far left). The lowest seam, the Bensham, was 858ft from the shaft top.

The St Hilda's mine was worked with the help of pit ponies, like many collieries in the North East. It featured extensive underground stables. Sacriston Colliery was the last mine in County Durham to use ponies. They were retired in 1985 when the mine closed. The last pit ponies in Northumberland were retired from Ellington Colliery in 1994.

Also pictured is a St Hilda's contraband warning notice which has survived. The risk of explosions meant that items such as cigarettes and matches were banned from coal mines.

Capped top of Crown shaft, Westoe Colliery:

A huge concrete cap on the Crown Shaft of Westoe Colliery at South Shields. This shaft was mainly used for coal-drawing and featured a very large skip. The seams worked included the Bensham, Hutton, Yard, Five Quarter, Brass Thill, Main and Maudlin. The other shaft of the mine, known as the Westoe, concentrated on man-riding. Today, the site of the colliery is occupied by a housing estate, known as Westoe Crown Village.

Eden Colliery main vent and fan house:

The impressive main vent and fan house of Eden Colliery, Leadgate, near Consett, County Durham. The introduction of powerful fans, at first driven by steam and then by electric power, was a major step forward for safety in the mines.

These fans helped to prevent any build-up of methane gas, or firedamp as it was known to the miners, and expelled the gas from the pit. They also kept the airways clear of any build-up of blackdamp, also known as chokedamp, in which there were very low levels of oxygen and very high levels of carbon dioxide.

Eden Colliery had two shafts, the Eden and the Stony. This picture shows the fan house at the Stony shaft. The workings below were developed as the Stony Heap Colliery, following the run-down of the Eden workings. A drift mine was also situated in the vicinity. The earliest colliery had opened in 1844. Final closure came in 1980. Other mines near Consett included Crookhall, Medomsley and Derwent collieries.

Wearmouth Colliery mine-gas vents:

Two mine-gas vents covered in wire mesh in sculptural form outside the Stadium of Light in Sunderland, which was built on the site of Wearmouth Colliery. In December 1993, Wearmouth, or Monkwearmouth, colliery was the last mine to close in the Durham Coalfield. It was also the deepest in the North-East, reaching down to around 1,800ft and, like Westoe, extended several miles out under the North Sea.

Friar's Goose pumping engine house:

The evocative ruins of the Friar's Goose pumping engine house on the banks of the Tyne near Gateshead Stadium. During the 19th Century this building housed very powerful machinery which not only drained the Friar's Goose Colliery workings, but also water from several other nearby mines both north and south of the river.

However, because financial backing was withdrawn this successful pumping engine ceased operating in 1851 and within less than 10 years nearly all the collieries involved were flooded. This ruin is now one of the oldest traces of the Great Northern Coalfield.

Weetslade pit heap:

Pictured in the background is the landscaped spoil or slag heap at the site of Weetslade Colliery, near Wideopen, a short distance to the north of Newcastle. This heap has been landscaped as a green hill and the colliery area is now a country park.

An artwork comprising three stone and metal columns crowns the summit of the landscaped hill, commemorating the mine and celebrating the site's new status as a country park.

Examples of two other landscaped spoil heaps can be seen at the site of the Rising Sun Colliery, in northern Wallsend. This area is also a country park. In addition, landscaping has taken place at the site of a large spoil heap close to the location of Fenwick Colliery, near Earsdon and Backworth.

Spoil heaps consisted of shale, stones, small particles of coal and dust. Shale is a slate-like rock which frequently had to be extracted to reach the coal. Many pit heaps in the North East were removed and the spoil was sometimes used as a foundation for new roads, the material being spread on the ground to create an even base surface. Other uses to which the spoil was put included infilling at former opencast sites, land reclamation in The Netherlands and ground preparation for the building of the MetroCentre at Gateshead.

Ashington colliery housing rows:
Former colliery housing rows at Ashington, near the site of Ashington Colliery. Wansbeck Business Park now occupies the land on which the colliery stood. Many former pit housing rows survive in the town, near the business park and in the Hirst area. The Ashington mine closed in 1986.

Haswell winding engine house:
The winding engine house of Haswell Colliery in County Durham. It is an impressive ruin, built mainly of stone with some brick details. This is probably the oldest surviving winding engine house in the Great Northern Coalfield. In front of the ruin is a sculpture commemorating the 95 men and boys who died as the result of an explosion at the Little Pit of Haswell Colliery in 1844. The ruin and the memorial can be found off Mazine Terrace, Haswell Plough.

The restored Blackfell hauler house.

Blackfell hauler house, Eighton Banks:

Some of the North East's coal railways also featured stationary engines which operated cable rope haulage along stretches of their routes, often in combination with the gravity of inclined planes. The Bowes Railway, part of which survives in working order in the Springwell area, used this system, as well as locomotives along some sections.

The Bowes Railway, which ran to staiths at Jarrow, opened in 1826 and was later extended to Kibblesworth and finally to Pontop in West Durham. It served several collieries.

The Blackfell Hauler House, situated on the route at Eighton Banks, survives as a reminder of the cable haulage system which enabled coal trucks to be moved on steep gradients. The building contained a stationary engine and cable drum. The drum still survives inside. The hauler house has undergone restoration work in recent years. It had suffered from vandalism. The restoration project was led by Tyne and Wear Building Preservation Trust in co-operation with the Bowes Railway Trust and Gateshead Council. Mount Moor Colliery was located not far from the hauler house and was linked to the railway.

An old level crossing gate on the route of the Bowes Railway at Eighton Banks.

Guards Wood Waggonway:

A clearly visible waggonway at Guards Wood, not far from Greenside in the Durham coalfield. The waggonway ran from near Prudhoe to Stella on the banks of the Tyne.

Waggonways, laid with wooden rail and operated by horses, began to appear in the North East in the 17th century. The first are believed to have opened around 1605-1608 near the River Blyth in Northumberland. By the late 18th Century a large number of waggonways had been created to the take the coal from the pits to staiths on the rivers Tyne, Wear and Blyth.

Iron rails first appeared on a North East waggonway in the 1790s. By the 1840s nearly all the region's coal railways, as surviving waggonways were by then called, had been converted to this more durable and efficient material.

The use of iron rails went hand in hand with the development of the steam locomotive, which was honed into workable form in the region by pioneering men such as George Stephenson, William Hedley, Timothy Hackworth and William Chapman. The North East was truly the 'Cradle of the Railways', born out of the need to transport coal as cheaply and efficiently as possible to the staiths.

Memorials to the miners

Pit wheel at Ashington:
A large pit wheel on Rotary Parkway, Ashington, which commemorates the town's mining heritage. It was unveiled in 2017 to celebrate the 150th anniversary of the sinking of Ashington Colliery's first shaft, the Bothal, in 1867. More than 300 Ashington miners lost their lives and the wheel is also a memorial to them. It is situated close to a number of former pit housing rows at the junction of Rotary Parkway and Booths Road. Among those present at the unveiling were Wansbeck MP and former president of the National Union of Mineworkers, Ian Lavery. Also attending was former Ashington Colliery pitman William Mason, aged 100. Ashington Colliery, a large mine which in its heyday employed thousands of pitmen, closed in 1988. Its site is now occupied by Wansbeck Business Park.

Village coal tub:
A coal tub on a green at Bowburn is a straightforward but highly effective memorial to the village's colliery and its miners. Tubs are among the most common of the memorials to be found at the sites of pits in Northumberland and County Durham. The pithead winding wheel is also a favourite commemorative choice.

One of the Bowburn miners' lodge banners carries a painting of Nurse Edith Cavell, who helped Allied prisoner of war soldiers to escape from occupied Belgium during the First World War and was executed by a German firing squad. The restored banner, which dates to c.1919, is unusual in carrying the image of a real-life woman, although others depict symbolic female figures. Mining was very much a male occupation. The pitmen of Bowburn thus honoured the brave and humanitarian Nurse Cavell.

Memories linger:
A full pit wheel commemorates Thornley Colliery, County Durham. It is located by the main road through the village, not far from the sight of the mine's Number 2 Shaft. An inscription states that Thornley Colliery was sunk in 1835 and closed in 1970. A further inscription includes the words: "Time passes, memories linger."

Marra at Horden:

This sculpture, entitled Marra, depicts a coal miner with his heart ripped out and stands in the colliery village of Horden, County Durham. Marra was created by sculptor Ray Lonsdale, of County Durham, and symbolises the devastation of the coal mining industry by the pit closures of the 1980s. Horden Colliery closed in 1987 following the Great Strike of 1984-85. The steel statue was unveiled in 2015 by a group of Horden's oldest surviving pitmen. The statue is situated in Horden Welfare Park. 'Marra' was the term used by North East miners to describe their workmates in the pits and was also a term of warm regard for any fellow miners and friends.

A memorial pit wheel at Horden overlooking the sea.

Coalface archer:

An artwork on a roundabout at Pegswood, which celebrates the mining heritage of south east Northumberland. Created by sculptor Tom Maley, it symbolically depicts a miner firing his shovel at the coalface, with a metal girder shaped like a bow. The work powerfully conveys the strength and vigour of the Northumberland pitman. It was commissioned by the Welbeck Estates, which has land in Nottinghamshire, including part of Sherwood Forest, and in Northumberland. The work is appropriately called Robin of Pegswood. Pegswood Colliery was sited only a short distance from where the artwork stands. The colliery closed in 1969.

Last North East deep mine:

This statue commemorates the centenary of the sinking of the first shaft of Ellington Colliery, last of the North East's deep mines. The impressive figure of a muscular pitman, complete with cap lamp and safety lamp, stands next to a winding wheel. It is situated at the entrance to the former mine, which became known as the 'Big E'. The statue is close to the Ellington Miners' Institute, now known as the Ellington Welfare Centre.

The artwork was created by sculptor Tom Maley, of Morpeth, and unveiled in 2009. Ellington Colliery closed in early 2005, ending hundreds of years of deep mining in the North East.

The Ellington miners' banner carries a highly original design. "Close the door on past dreariness," declares the message on one side of the standard, and on the other side the motto reads: "Open it to future brightness." The words are accompanied by paintings of dreary pithead housing and by contrasting pleasant, new housing in a tree-lined street. The banner was designed by Oliver Kilbourn, a famed member of the Ashington Group of pitmen painters. The group is the subject of Lee Hall's acclaimed play, The Pitmen Painters.

Pit Disaster Memorials

New Hartley disaster memorials:

Pictured right, the memorial to the 204 men and boys who died in the New Hartley pit disaster of 1862. It stands in St Alban's Churchyard at Earsdon, between Shiremoor and West Monkseaton, North Tyneside. The death toll was the largest of all the North-East pit accidents.

Several boys lost were aged only 10. Unusually, the tragedy resulted from part of a cast iron pumping engine beam falling down the pit's single shaft, carrying with it a mass of debris, and thus blocking the only means of escape for the miners working below.

The faces of the memorial are inscribed with the names of all those who died. There were few if any homes in the village of New Hartley to escape a bereavement and in some cases families suffered multiple bereavements. One older miner perished with three of his sons and a grandson.

The New Hartley disaster led Parliament to pass an Act which stipulated that all mines should have at least two shafts or other exits so that an alternative means of escape was possible.

It also led to the setting up of the Northumberland and Durham Miners' Permanent Relief Fund, which provided payments to the widows and families of men killed in pit accidents, on the condition that the men concerned had contributed to the fund from their wages. The fund also provided payments to miners injured as a result of their work and later introduced small pension payments for retired miners.

The memorial is sited to the rear of St Alban's Church. The church features two stained glass windows, above a balcony area, which commemorate those who lost their lives in the disaster.

New Hartley memorial garden:

Four miles away from Earsdon, in the village of New Hartley itself, a small memorial garden, not far from the railway level crossing on the eastern edge of New Hartley, has been laid out at the site of the mine, also known as the Hester Pit. Pictured (top far right) is the capped shaft - a large, rectangular shaped block of stones - which is situated to the rear of the garden. To the left, is a circular stone structure marking the small shaft which held some of the upper pump rods, but which did not link up with the deeper sections of the mine.

Commemorative paving stones leading to the site of the Hester pit shaft feature the names of the 204 men and boys who died and moving words telling of the tragedy. Flowers are depicted on these stones. The artwork on the pathway is by artist Russ Coleman and the words by writer Rob Walton, working in co-operation with New Hartley community groups and local schools. Members of the team involved in producing the path included Leigh Cameron, Phill Blood, Kirk Teasdale and Blyth Valley Arts. Our other pictures show close-ups of two of the paving stones.

In addition, a memorial banner was produced to mark the 150th anniversary of the tragedy. The banner, created in duplicate, was designed by artist Alison Walton-Robson and has been on display in the village.

16th January 1862

Counted 204

The tally lad who knew them all

John Dawson, 12   Johnson Sharp, 13

William Davidson, 11   Thomas Sharp, 48

John Davidson, 38   Patrick Sherlock, 28

Philip Cross, 20   George Skinner, 14

The tally lad who knew them all

Counted 204

New Hartley memorial windows:

Pictured are the two New Hartluy disaster stained glass memorial windows in St Alban's Church at Earsdon. The windows commemorate the 204 men and boys who died in the disaster of 1862. They were created by artist Cate Watkinson, who is based in Newcastle and specialises in glass and public art.

One window features images representing a single shaft, the coal seam and points of light symbolising the miners who lost their lives. The other, more brightly coloured window suggests the hope engendered by the passing of safety legislation following the tragedy. The windows are positioned above a balcony area of the church.

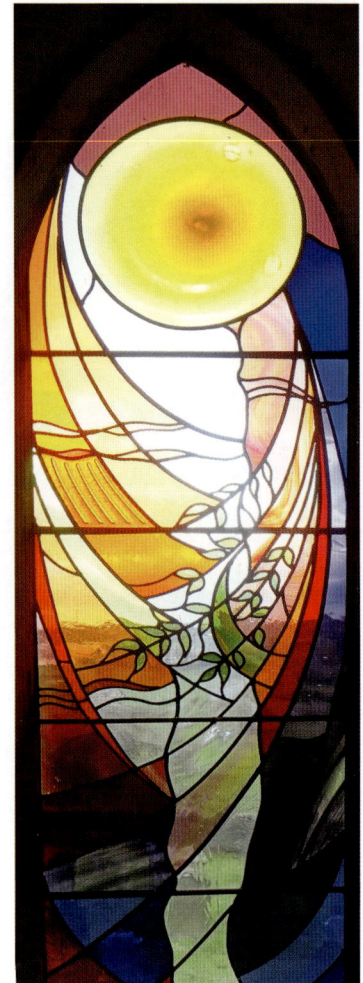

For the darkness is as light with Thee

West Stanley disaster memorials:

The main memorial to the second worst North East pit disaster is located at the eastern end of the High Street in Stanley, County Durham, close to Slaidburn Road. It commemorates the 168 men and boys who died as the result of an explosion at West Stanley Colliery, also known as the Burns Pit, in 1909. Fifty nine of the 168 victims were aged under 20.

An extensive search failed to recover the bodies of two miners. These were not found until 24 years later when they were revealed by a roof fall. Exactly how the blast - believed to be a coal dust explosion - was ignited remains a mystery.

The main West Stanley memorial features two impressive halves of a pit winding wheel in parallel with one another (pictured right) and a wall bearing the names of the men and boys who lost their lives, together with the names of the seams of coal they were working in on the day of the tragedy. In front, a fine mosaic laid on the ground includes images of a safety lamp, miner's shovel and pick. It is the work of artist Anne Brady.

The memorial was unveiled in 1995 by former Newcastle United manager Kevin Keegan, a grandson of one of the rescuers, Frank Keegan. Frank had been working in the pit when the disaster happened, but after being rescued returned in a bid to save his comrades. A plaque makes clear that the memorial is 'also dedicated to all miners and their communities'.

In 2009, the centenary year of the disaster, a remembrance service for those who died was held at St Andrew's Church, Stanley. A second service took place at the memorial on the exact anniversary of the tragedy, February 16, when the bells of St Andrew's were rung a symbolic 168 times.

An earlier memorial (inset) to the victims of the West Stanley tragedy can be found inside the entrance to the cemetery at East Parade, Stanley. It is an ornamented pink and grey marble pillar and was erected in 1913 by social clubs of the district.

A large number of those who died are buried in St Andrew's Churchyard where a memorial was erected in 2005. Others who lost their lives in the disaster were interred in St Joseph's Churchyard, Stanley, where there is also a memorial stone.

The West Stanley disaster eventually resulted in the introduction of the identity disc system in Britain's collieries. Under this rule each man left a numbered disc on the surface before descending into the mine. This enabled the identities and number of men working underground to be known should an accident occur or men go missing.

The West Stanley miners' banner depicts a pitman in workings hit by an explosion, with the words "The unknown miner." Thus a parallel is drawn between the dangers faced by miners and those faced by soldiers.

Seaham disaster memorials:

In 1880, an explosion at Seaham Colliery, County Durham, resulted in the deaths of 164 men and boys. The exact cause of the blast became the subject of dispute.

Our top photograph shows the memorial to the tragic 164, which can be found in the grounds of Christ Church, Seaham, opposite the site of the now closed colliery.

The names of those who lost their lives are inscribed in gold-coloured lettering on the face of this tribute in stone, surmounted by a cross. One of the inscriptions reads: 'What man is he that liveth and shall not see death.'

The memorial is sited within a small garden of remembrance, which also features a second tribute - a stone column, also topped by a cross (pictured below left). It commemorates the 26 men and boys who died in an explosion at Seaham Colliery in 1871. An inscription at its base states that the memorial was erected by their fellow workmen and others in 'loving memory' of the 26.

St Hilda's Colliery disaster memorial:

Pictured below: A pithead winding wheel commemorates the miners who died in the St Hilda's Colliery disaster at South Shields in 1839. The workings had until then been considered safe from a gas explosion. However, 51 men and boys died as a result of a firedamp explosion.

The memorial is sited next to the Asda store in South Shields and close to the surviving St Hilda's pithead building. Most of the dead were buried in St Hilda's Churchyard.

Wallsend disaster memorial plaque:

The fourth North East pit disaster in which the death toll reached treble figures occured at Wallsend Colliery in 1835. A total of 102 miners were killed as the result of an explosion.

The men and boys were buried in a mass grave in St Peter's Churchyard, Wallsend. Almost incredibly, it seems the grave was unmarked. Certainly no such marker survived, although it is possible a temporary tribute, perhaps featuring a wooden structure, was erected and later lost or destroyed by the elements.

However, nearly 160 years later, in 1994, a plaque (pictured) was unveiled on the inner south wall of the churchyard, thanks to the efforts of Wallsend Local History Society, Northumberland NUM and North Tyneside Council. The inscription makes clear that the youngest miner was aged eight and the oldest 76.

Among the dead was a man surrounded by a group of boys. He had evidently been trying to lead the boys to safety towards the nearest shaft, the G Pit, although had they succeeded in reaching it they would have found it blocked by the explosion. Poignantly, each lad was discovered with his cap in his mouth in a bid to combat the effects of the afterdamp.

Haswell disaster memorial:

Pictured is the memorial to the victims of the Haswell Colliery disaster, located at Haswell Plough, a few miles west of Easington in County Durham. It takes the form of a sculpture sited in front of the ruins of the Haswell Colliery winding engine house. The memorial commemorates the 95 men and boys killed when an explosion engulfed Haswell's Little Pit in 1844. Only four miners survived.

The sculpture takes the form of a relief in the lower part of a large white stone and depicts the faces of men, women and children in a coal seam or mine tunnel. The work, by sculptor Michael Disley, is entitled Spirit of Haswell. Part of the inscription reads: 'Their spirit lives on'. Each of the 95 railings fronting the memorial carries a face, symbolising those who lost their lives. The ruins of the winding engine house, in stone and with some brickwork, are a remarkable survival.

Felling disaster memorial:

Pictured right is the small obelisk memorial to 91 men and boys who died in the Felling pit disaster of 1812. It is located in St Mary's Churchyard, Heworth, Gateshead, across the road from Heworth Metro Station. Recovery of the bodies took around four months and the 92nd victim was never found. The youngest of those who died were two eight-year-old boys, who operated ventilation doors in the mine. Typically, the disaster was caused by a firedamp (methane gas) explosion.

The modest size of the memorial seems at odds with the magnitude of the disaster. However, recently a commemorative plaque has been unveiled on the wall below the obelisk, a further tribute to the victims.

The Rev John Hodgson, of Heworth, conducted the funeral service for the 92 victims and afterwards wrote a detailed account of the tragedy and its aftermath. His description of the event led to the establishment of the Society for the Prevention of Accidents in Mines, based in Sunderland.

The society was able to call upon the services of scientist Sir Humphry Davy to look into the possibility of developing a safety lamp which it was hoped would provide a light without the potential for explosions.

Davy visited the North East to carry out experiments for his lamp at about the same time as North-East-born steam locomotive pioneer George Stephenson was also experimenting, at Killingworth Colliery, to devise his own safety lamp, known as the 'Geordie'. The two lamps were similar in principle, although not in form.

Dr William Clanny, a Sunderland physician, had already invented a safety lamp, but it proved to be impractical. However, Clanny went on to develop four more versions of his lamp, each one an improvement on the other and which were far more effective.

Despite all these developments, the early safety lamps were by no means perfect. Yet they proved to be superb gas detectors and it was in this role that they excelled, warning when methane was present.

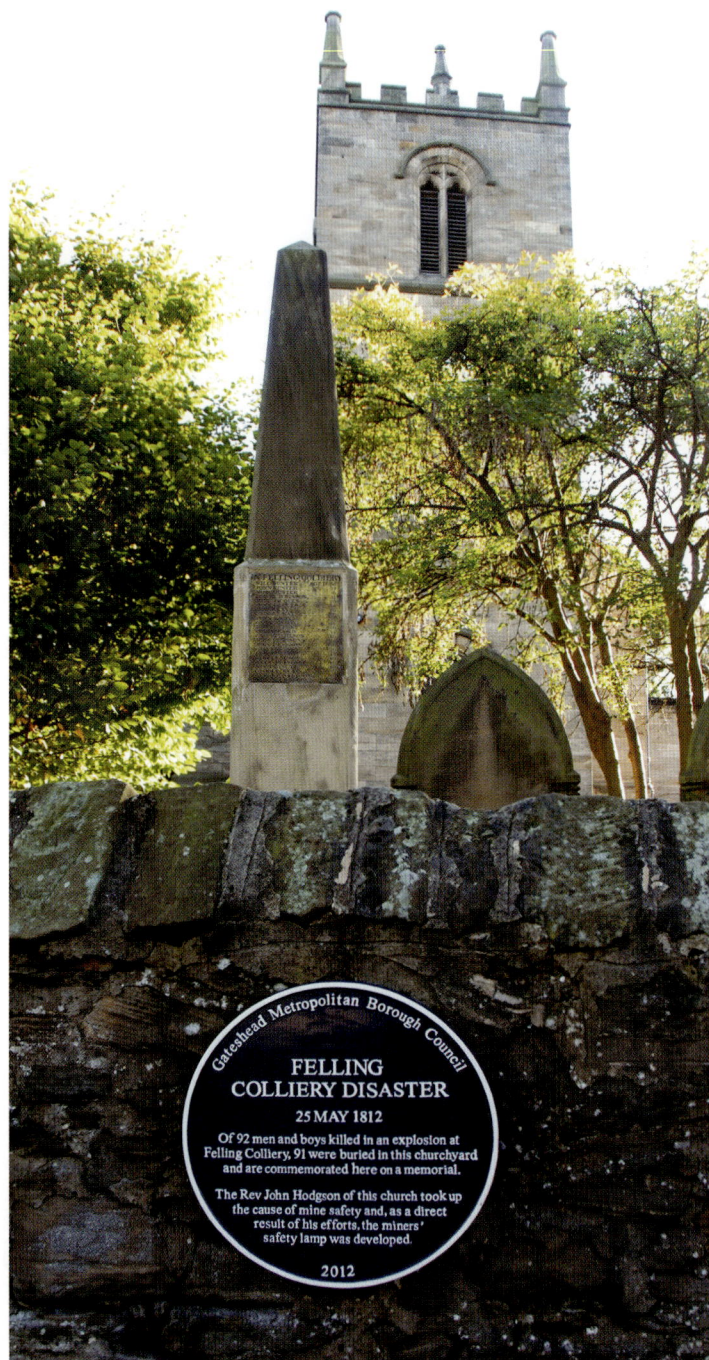

FELLING COLLIERY DISASTER
25 MAY 1812
Of 92 men and boys killed in an explosion at Felling Colliery, 91 were buried in this churchyard and are commemorated here on a memorial.

The Rev John Hodgson of this church took up the cause of mine safety and, as a direct result of his efforts, the miners' safety lamp was developed.

2012

Trimdon Grange and Tudhoe disaster memorials:

In February 1882, an explosion at Trimdon Grange Colliery in County Durham resulted in the deaths of 74 miners. The Trimdon Grange workings were linked to those of Kelloe (East Hetton) Colliery and afterdamp formed in the wake of the blast spread to Kelloe, killing six men there.

One memorial to the Trimdon Grange victims is situated in a cemetery on the eastern edge of Trimdon Village. Forty four of the 74 are buried here. This tribute in stone features relief carvings of a miner being rescued from a pit, a grieving widow and daughter at a graveside, a pitman making his way to work and clasped hands symbolising friendship. Another tribute to those who lost their lives, a memorial pit wheel, can be found in Trimdon Grange.

Twenty six of the dead are buried in Kelloe Cemetery at Church Kelloe, where there is a memorial of identical design, including the carvings. Miners from the Kelloe area were among the victims. The inscription records that as well as the 26 at Kelloe, one miner is interred at Croxdale, one at Cassop cum Quarrington and two at Shadforth.

Only two months after the Trimdon Grange explosion, a similar tragedy hit Tudhoe Colliery in the Spennymoor district. The blast, in April 1882, resulted in the deaths of 37 miners. A memorial to the Tudhoe dead was erected in York Hill Road Cemetery, Spennymoor, and is identical to those at Trimdon Grange and Kelloe. The photographs show the Trimdon Village memorial and carvings on one of the three memorials.

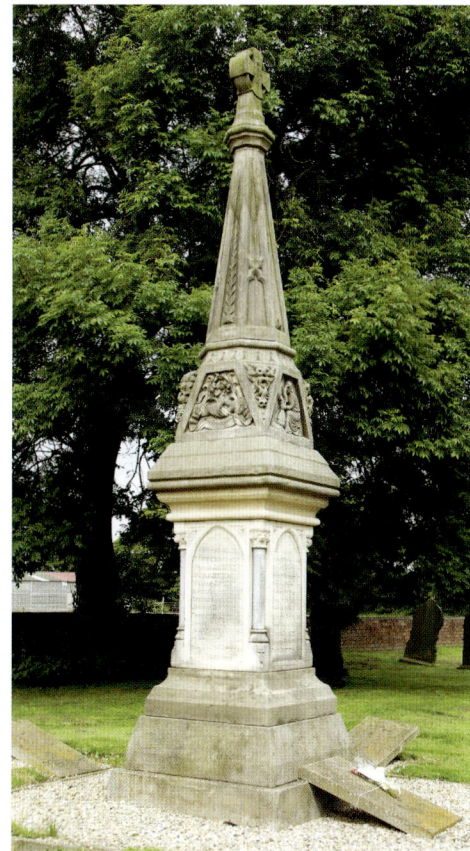

Easington Colliery disaster memorials:

The last of the major tragedies to hit the Great Northern Coalfield occurred at Easington Colliery, County Durham, on May 29, 1951. A firedamp explosion, spread by coal dust, engulfed a district of the mine and 81 pitmen were killed. Two rescue workers, who were also miners, lost their lives in the resulting afterdamp. An 18-year-old pitman was brought out of the mine alive but died in hospital shortly afterwards.

The chief mines inspector found that the explosion had been caused by a coal cutter machine coming into contact with pyrites, thus producing sparks which ignited the blast.

Pictured are views of the main memorial to the victims in Easington Colliery Cemetery. Seventy two of the dead are buried side by side in a memorial garden at the centre of the cemetery, reached through an avenue of trees.

At one end of the garden is a relief sculpture of a miner, complete with helmet and carrying a safety lamp. Next to him is a large lump of coal surmounted by a cross. Flanking the relief are two coal-filled tubs on rails. An inscription reads: 'Remember before God those who gave their lives in the Easington Colliery disaster in May Nineteen Hundred and Fifty One to whom this garden of remembrance is dedicated.'

Close to the relief is a coal-cutter machine on rails, a poignant reminder of the cause of the disaster. At the other end of the garden, a relief facing the graves depicts safety lamps. The garden of remembrance also features sculptures of safety lamps and miners' picks arranged as crosses.

Across the main road through the village, an avenue of trees leading to Easington Colliery Welfare Park also commemorates those who died, each tree representing one of the victims. The first tree was planted by a 16-year-old Easington miner in 1952. In the avenue a large stone from the scene of the tragedy bears a memorial tablet urging passers-by to 'get understanding and promote goodwill in all things'.

A more general memorial to all those who worked at the colliery, above and below ground and including all those who died during its lifetime, can be found at the site of the mine, close to the sea. It takes the form of a community garden and is surrounded by attractive metal railings. Its gate replicates images on an Easington miners' union lodge banner, including the faces of pitmen's leaders Thomas Hepburn and Lawrence Daly.

The garden features a pit wheel arranged horizontally with plants growing around and between the spokes. Medallions with various motifs, created by schoolchildren, are set in paving stones in front of the wheel.

Nearby, on the side of a mound overlooking the sea, is a tall, three-deck pit cage from the mine's South Shaft - another memorial to the colliery and its pitmen. The capped shaft is only yards way, on top of the mound. Easington Colliery closed in 1993.

Hope for the future:

A statue in Scotswood, Newcastle, commemorates the area's coal mining history and the 38 pitmen and boys who died in the Montagu View pit disaster of 1925. This artwork depicts a pitman and his pony with a boy and girl riding on the pony's back. Entitled Yesterday, Today, Forever, the miner is leading the youngsters into a better future. The message is one of hope.

The statue was made by Xceptional Designs and officially unveiled in 2013. It was funded by Newcastle City Council's Make Your Mark project. The work stands on Scotswood's Fowberry Road roundabout. The Montagu View pit disaster was caused by flooding.

Affectionate regard:

Pictured is the memorial to the 28 pitmen who died as the result of an explosion at Elemore Colliery in 1886. The blast is likely to have been caused by shot-firing ingniting coal dust.

The memorial can be found in St Michael and All Angels Churchyard at Easington Lane, near Hetton Lyons. Nearly all the gravestones in the churchyard are those of the victims.

An inscription on the base of the column, which is surmounted by a cross, carries a quotation from the Bible: "There is but a step between me and death." The main plaque states that the memorial was erected to the memory of the 28 "by their fellow workmen, employers and friends as a mark of affectionate regard".

The names of these tragic Elemore miners, some of whom were teenagers, are inscribed on the faces of the base, although the effects of weather have rendered one face difficult to read.

Let them rest from their labours:

The obelisk in Wingate village, County Durham, which commemorates the 26 miners who lost their lives in the Wingate Grange Colliery disaster of 1906. It was found that the firedamp explosion had been caused by shot-firing.

This sandstone memorial, restored in recent years, bears the names of the dead, and the fitting inscription: "Let them rest from their labours."

Walker Colliery tragedy:

The memorial to eight men who lost their lives as the result of the explosion at Walker Colliery, Newcastle, in 1887. It stands in Walker Cemetery and takes the form of a large stone pillar surmounted by rounded ornamentation. The inscription makes clear that the tribute was erected by "officials and workmen" of the mine.

Below the names of the eight pitmen who died, is that of another miner killed in a similar accident at Walker Colliery barely two months earlier. Other accidents at the mine included a disaster of 1862 in which 16 miners lost their lives as the result of an explosion. The pit closed in 1920.

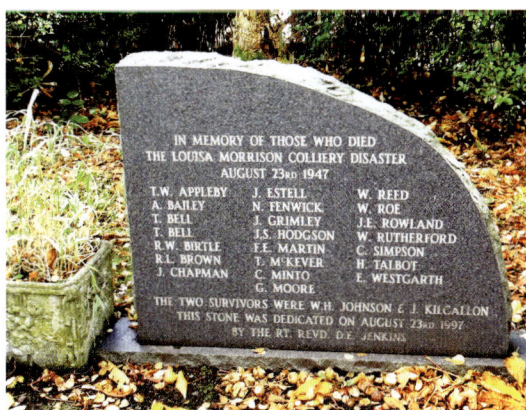

Louisa Morrison Colliery disaster memorial stone:

Pictured left, this memorial stone is for 22 miners who died in an explosion at Louisa Morrison Colliery, Stanley, in 1947. It is located in St Aidan's Churchyard, Annfield Plain. The blast was the result of a lethal combination of firedamp and coal dust. Ignition was believed to have been caused by the striking of a match. The youngest pitman lost was aged 18.

Tragedy at Woodhorn:

The statue on the memorial to the 13 miners killed in a firedamp explosion at Woodhorn Colliery, Northumberland, in 1916. On the day of the disaster a ventilation fan was working at a slower speed than usual and an air compressor was off. The Woodhorn mine had previously been regarded as normally free of methane.

The year 2008 saw the re-dedication of the memorial. The white obelisk, designed as a drinking fountain, is surmounted by the fine statue of a pitman holding a safety lamp and pick. The miner, a deputy, is shown holding the lamp up to test for the presence of gas. This work was created by sculptor John Reid.

For many years the statue was sited in Hirst Park, Ashington, but in 1991 was moved to the car park of the Woodhorn Colliery museum. The re-dedication service took place after it was again moved, this time to a location near one of the museum's two winding wheel towers.

Eleven of the 13 men who died in the Woodhorn disaster were married with 34 children between them. After the tragedy, only flame safety lamps were allowed on the coalfaces at Woodhorn until the advent of the electric cap lamp. The carbide lamp, a naked flame lamp, was banned, although it was used in many other pits in Northumberland which were regarded as largely free from methane gas. Deputies, however, were still required to test for gas at the start of each shift in all collieries.

Nine remembered:

A full pit wheel commemorates the miners of Eppleton Colliery, including nine who died in a disaster at the pit in 1951. In 2011, a plaque in memory of the tragic nine was unveiled at the wheel by Dorothy Robinson, whose grandfather was among the victims, and Anne Phillips, whose father also lost his life in the disaster. The tragedy was caused by an explosion. The winding wheel is from the mine and is located in Hetton Lyons Country Park, Hetton-le-Hole. Eppleton Colliery opened in 1833 and closed in 1986.

Youngest life lost:

Olive Tindle stands beside the grave of her relative, Thomas Agar, who was killed in the Glebe Colliery explosion at Washington in 1908, along with 13 other miners. Thomas, aged 18, was the youngest victim.

Olive has regularly visited Thomas' grave in the cemetery behind Holy Trinity Church in Washington Village and lays flowers there. The grave of several other victims of the tragedy can be found nearby, not far from the gates. This photograph was taken in 2011.

A parchment tribute to the 14 miners who died is displayed in a glass-fronted case at Holy Trinity Church, next to Washington Old Hall. The Glebe disaster was caused by a combination of shot-firing and firedamp.

Montagu Pit disaster:

The Montagu pit disaster memorial in St John's Cemetery, Elswick, Newcastle. The tragedy, which happened on March 30, 1925, was the worst in the Great Northern Coalfield between the two world wars.

The Montagu View pit, on the western edge of Scotswood, Newcastle, was engulfed by an inrush of water from old mine workings. They turned out to be those of the former Paradise Pit, abandoned in 1848. The sudden flood led to the deaths of 38 men and boys. Of these, 21 were drowned and 17 probably succumbed to chokedamp.

The tragedy might have been avoided had those in charge been given access to plans of the abandoned pit workings, which it was later discovered did exist. Author A.J. Cronin included scenes based on the Montagu pit disaster in his novel, The Stars Look Down.

The memorial features statues of a miner with safety lamp and pick and of the Good Shepherd.

In 2006, a memorial garden, with a pavement area on the theme of a pithead wheel, was created in the grounds of St Margaret's Church, Scotswood.

In 1925, the same year as the Montagu disaster, five men lost their lives as the result of an explosion at the Edward Pit, on the northern edge of Wallsend.

Lethal combination:

In 1822, 38 miners died at Stargate Colliery, near Ryton, as the result of an explosion. Four of the dead, including a boy, were blown from the shaft by the force of the blast. The miners were about to start a working shift when the tragedy happened. An explosive combination of candles and gas was thought to be the cause.

Most of the 38 were buried in Holy Cross Churchyard, Ryton, but their graves appear to have been unmarked. However, in 1993 a stone memorial tablet to the dead was placed at the approximate site of the graves by Ryton Heritage Group.

In 2018, a memorial coal tub (pictured) was unveiled to these tragic men and boys. Located in Stargate Lane, it is fronted by a plaque listing those who lost their lives. Two of the boys were aged 10. The Stargate pit disaster left 19 widows and 62 children. Nearby are three former colliery housing rows – Low Row, Middle Row and High Row.

Coal dust danger:

In 1885, Usworth Colliery, north of Washington Village, was the scene of an explosion which killed 42 men and boys. Pictured (right) is the memorial column to the dead in Holy Trinity Churchyard at High Usworth, Donwell, Washington. A lethal combination of shot-firing and coal dust was found to be the cause.

Most of the victims are buried in the Donwell churchyard, but Roman Catholic miners who died are interred at the Church of Our Lady, Washington Village. Their memorial (pictured far right) is surmounted by a Celtic cross.

The names of all those who died, Catholic and Protestant, are inscribed on the faces of both memorials, including two boys aged 14.

Explosion and flooding:

A memorial obelisk in St Bartholemew's Churchyard, Thornley, is to miners killed in various accidents at pits in the area. The base of the memorial states that it was erected by workmen of the Ludworth, Thornley and Wheatley Hill collieries in memory of four of their fellow miners who died as the result of a gas explosion in 1876.

Two sides bear inscriptions to other miners, including those lost in an accident at Thornley Colliery in 1878 and as the result of flooding at the Wheatley Hill pit in 1871. The youngest miner listed is aged 13.

It should be emphasised that this memorial does not list all those killed at the three pits. The St Bartholemew's Church building has now been demolished.

Murton pit wheel memorial:

Pictured is the pit wheel memorial to all those who worked at Murton Colliery in County Durham, including those miners who lost their lives in disasters in 1848 and 1942. The 1942 tragedy, during the Second World War, was the result of an explosion. It claimed the lives of 13 pitmen. The wheel has now been moved from its original position next to a green area in Murton to a more central location in the village.

To all those lost:

A pit wheel and coal tub set amid flowers on a grass verge by the main street in Burradon, North Tyneside. This is a memorial to all miners who died at Burradon Colliery during its lifetime from 1820 to 1975 – including the 76 lost in a disaster of 1860. The wheel is from the mine. The disaster was caused by a firedamp explosion.

Tributes and Remembrance

Wheel on roundabout:

A winding wheel from Ellington Colliery on a roundabout at the eastern edge of Westerhope Village in Newcastle. This survivor from the region's last deep mine is a memorial to North Walbottle Colliery and its pitmen. North Walbottle Colliery at Westerhope was one of a cluster of mines on the western side of Newcastle. They included Blucher at Blucher Village, the Duke Pit at Walbottle, Throckley Maria and Throckley Isabella. The wheel was officially presented to the Westerhope community by former pitman Ian Lavery, at that time president of the NUM and now MP for Wansbeck.

Village gem:

An attractive "half" pit wheel, fronted by flowers, commemorates Kibblesworth Colliery, near Gateshead, and its miners. It is situated on a small green in the village of Kibblesworth. A road leads uphill from the memorial to the site of the mine, which closed in 1974.

Family group at bus station:

The fine statue of a miner, his wife and son, situated in a garden outside the bus station in Concord, Washington. The miner's son is depicted handing his father his bait tin.

This artwork, by sculptor Carl Payne, was unveiled in 2012 by Dave Hopper, general secretary of the Durham Miners' Association.

It was commissioned by Washington Miners and Community Heritage Group, which included pitmen who worked at Washington F, Glebe and Usworth collieries.

The sculpture pays tribute to the coal mining community of the Washington area and represents a pitman and his family stepping out into a brighter future.

Esh Winning memorial:

Another impressive statue, of a miner, his wife and small daughter, can be found at Esh Winning in County Durham. It was designed by Norman Emery, Durham Cathedral archaeologist in residence, and produced by monuments firm Odlings, of Hull.

This artwork was unveiled in 2015 and is located in the village's war memorial garden in Woodland Road and close to the communal hall. It is a memorial to the miners of Esh Winning and the mining communities of the Upper Deerness Valley. The mines of this area included Esh Winning, Waterhouses, Hedley Hill and East Hedleyhope.

The pitman, wearing knee pads, is depicted holding his daughter while his wife carries a book with the words "Shield the Oppressed." Also included in the work are the words: "Social justice, compassion, peace and liberty."

The Roundy at Ushaw Moor:

Pictured is the memorial to the miners of Ushaw Moor Colliery. It takes the form of a sculpture of a massive lump of coal. The work, by Colin Rose, is entitled The Roundy, a word used by pitmen to describe a very large lump of coal. The Roundy can be found on a green next to the main road through Ushaw Moor.

Above the shaft:

This large pit wheel at Albany, Washington was in use at Silksworth Colliery for many years. It is a tribute to the miners of Washington F Pit, which was one of the oldest collieries in County Durham. The wheel stands directly above an old shaft of Washington F, now capped. The man in the picture is former miner Jim Wilson. Also situated in the Washington area were the Glebe and Usworth collieries.

Men who gave their all:

Two large, "half" pit wheels in Middle Street, Blackhall, County Durham, which are a memorial to all those who worked at Blackhall Colliery, including those who lost their lives at the pit. A plaque indicates that the wheel was presented by Blackhall Settlement Renewal. It includes the words: "A small token of respect we bring for men who gave their all." Blackhall colliery closed in 1981.

Wearmouth lamp and pit wheel:

The giant miner's safety lamp outside the Stadium of Light, Sunderland. The stadium was built on the site of Wearmouth Colliery. The larger-than-life lamp, which stands in the centre of a roundabout near the ticket office, is the work of sculptor Jim Roberts and was manufactured by Armitage Engineering Ltd. It is kept permanently lit and commemorates the industrial heritage of the area, including coal mining. Also pictured is a pit wheel, erected on the other side of the stadium, overlooking the River Wear. It is a memorial to Wearmouth Colliery and its miners.

Miner at the coalface:

Massive slabs of rock cover the statue of a miner in a low-roofed seam at Fishburn, County Durham. The memorial, created by sculptor Keith Maddison, was presented to the community of Fishburn by the village's Millennium Group in memory of all mineworkers of the village colliery. It is sited on a green close to the main road junction in Fishburn. The miner is working with his pick while down in the extremely narrow confines of the seam. A bed of coal surrounds this highly original and prominent tribute. A winding wheel from the colliery is situated in a recreation field on the southern edge of the village.

Pitman in the seam:

Pictured, bottom right, is the impressive statue of a miner working at a coalface at the site of Lambton Colliery and Cokeworks, now a country park, between Bournmoor and Fencehouses. This artwork, which includes depictions of pit props, is by sculptor Colin Wilbourn. It is entitled Underground.

Colourful memorial:

Four coal tubs on rails at Jubilee Park, Spennymoor, County Durham, commemorate all miners who died in the pits of the Spennymoor area. Collieries in the surrounding area included Tudhoe, Whitworth Park, Dean and Chapter, Merrington and Thrislington.

In the middle of a bed of flowers fronting the tubs is a pit lamp in a glass case. A plaque states: "This pit lamp has a special light and always it will shine on the names of the miners – men and boys – who died and worked in the mine."

The memorial was opened by Spennymoor Town Council Mayor J.M. Marr in 2000. This tribute is a blaze of floral colour on a spring or summer day.

Pit lad with pony:

The bronze statue of a young miner with a pony and tub at Blyth in Northumberland. The pit lad – a driver or pony putter – is depicted sitting on the "limbers" which attach the animal to the tub. The artwork is situated in a small community garden on a housing estate off Cowpen Road, Blyth. The work was created by sculptor Richard Broderick.

A plaque states that the garden, created by local schoolchildren, is dedicated to the memory of all miners who spent their working lives in the pits of Blyth between 1794 and 1986. An inscription reads: "Our inspiration for a better future." The town was once the home of collieries such as Cowpen, Bates, Crofton Mill, Isabella and New Delaval. The Crofton Mill pit was close to the centre of Blyth.

Bates was the last of the town's collieries to close and a winding wheel from its manriding shaft stands near Broadway Circle in Blyth as a memorial to its miners.

Miners at work:

Pictured are two relief artworks showing miners at work by well-known North-East artist Bob Olley. They are fixed to the gate posts of the entrance to the former Whitburn Colliery, on the coast near South Shields. Bob, who was a miner at the colliery, presented the artworks to the people of Whitburn and they were unveiled by the Mayor of South Tyneside, Councillor Tracy Dixon, in 2008.

The artworks are in concrete, bronze and resin. The Whitburn mine, which extended out under the North Sea, closed in 1968. It was sometimes referred to as Marsden Colliery.

Pitman on the green:

A sculpture at Craghead, County Durham, depicting a pitman. The figure stands on a grass verge beside the main street of the village. The artwork, entitled Looking Out Looking In, portrays the miner looking through a doorway in the pit workings. An inscription states: "In memory of Craghead miners. Lest we forget. 1839-1969." The work was created by sculptor Jim Roberts. Further westwards along the road a large sculpture of a safety lamp (pictured) is another memorial to the pitmen of Craghead Colliery. This is also by Jim Roberts.

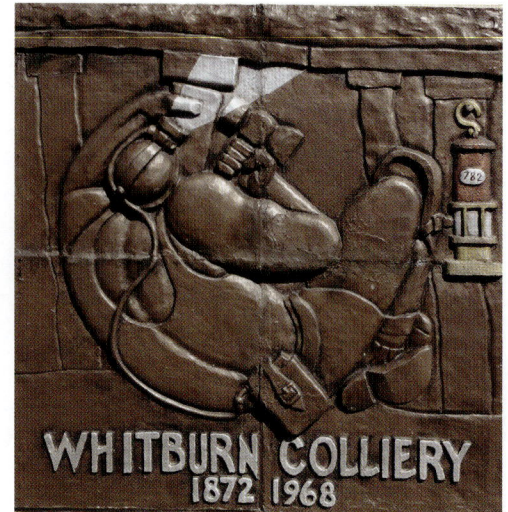

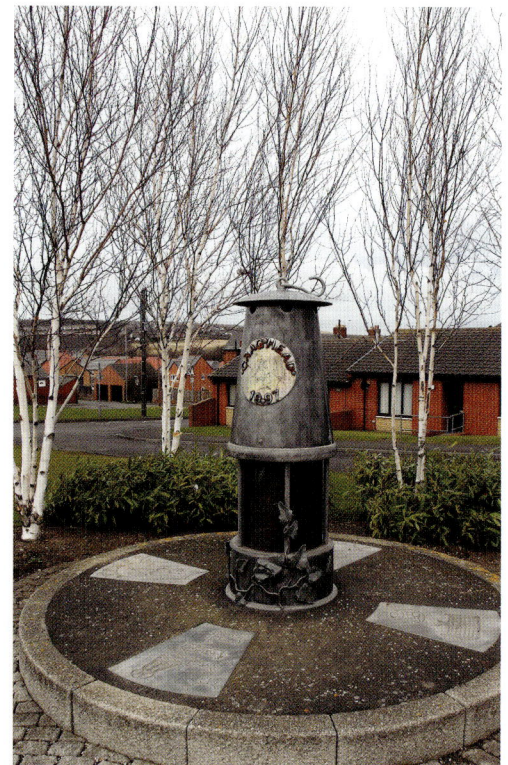

Mining brothers:

An artwork entitled The Brothers Waitin' t'gan Down depicts three miners. Situated on the seafront in Seaham town centre, it was created and designed by artist and former Silksworth miner Brian Brown. The work was unveiled in 2011 in memory of all who worked at Seaham, Dawdon and Vane Tempest collieries.

Impressive tribute:

Two large 'half'-winding wheels, and a smaller wheel, are a tribute to Wingate Grange Colliery and all those who worked in it. This impressive memorial stands next to a road junction in Wingate, a cable links the wheels, the smaller wheel represented as if on top of the pithead winding tower.

In 1906, Wingate Grange Colliery was the scene of a disaster in which 26 miners lost their lives. The mine closed in 1962.

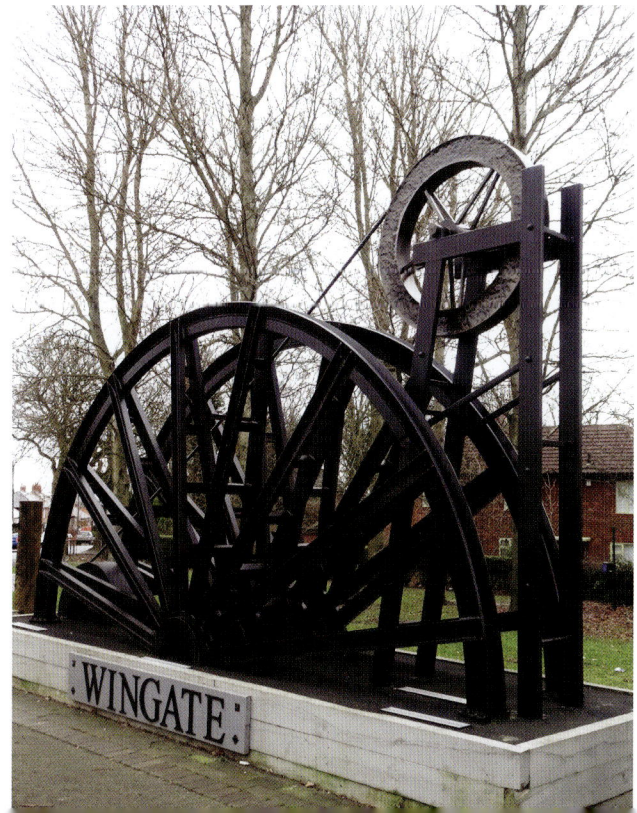

And the Village Remains: The Last Tub:

Pictured is the statue of a miner pushing a tub, which was unveiled in 2016 at South Hetton. It is a tribute to the mining community of South Hetton and was created by sculptor Ray Lonsdale. The artwork was unveiled by the first woman to work at South Hetton colliery, Joyce Raymond, and also present was Bill Short, aged 95, the oldest surviving miner from the colliery. The sculpture is entitled And the Village Remains: The Last Tub. South Hetton Colliery closed in 1962.

Ray Lonsdale's other works include the statue of a miner at Horden Welfare Park and the giant statue of a First World War soldier at Seaham.

Path to the seams:

Two 'half' pit wheels flank a short pathway, representing a shaft, leading to a symbolic representation in stone of the working places of miners in a coal seam. This tribute is located within the Miners' Memorial Garden at Herrington Country Park, which was laid out on the site of the former New Herrington Colliery. The memorial pit wheels are sited above the skip winding shaft of the mine.

Reminders of coal era

Causey Arch:

A reminder of the era when coal was taken by horse-operated waggonways to the staiths on the rivers Tyne and Wear for onward shipment to many destinations. Pictured is Causey Arch, near Stanley, County Durham. This impressive single-span bridge, which carried a waggonway over the dene of the Causey Burn, is believed to be the oldest surviving single-arch railway bridge in the world. It was built in 1725-1726. The bridge linked Tanfield Colliery with staiths on the Tyne.

Philadelphia power station:

Philadelphia Power Station, near Shiney Row, built in c.1906 for the Durham Collieries Power Company. Its purpose was to supply electricity to various mines in the area. By 1911 the station had been incorporated into the Newcastle upon Tyne Electric Supply Company network. It was later used as a garage by the National Coal Board.

Whinfield coke ovens:

Beehive coke ovens at Whinfield, near Rowlands Gill, which were the last of their type to be used in Britain. The ovens - so called because of their beehive-shaped domes - were operated from 1861 to 1958. Five out of the original 193 at Whinfield have been saved for posterity. When coal was coked (ie baked) a number of elements were removed in the process, notably sulphur. The resulting coke, a smokeless fuel which burns hotter than coal, was ideal for a number of industrial processes, including steel-making and malting.

Inkerman coke ovens:

Beehive coke ovens at Inkerman, near Tow Law, County Durham. Coal for the ovens, which date to the 19th Century, came from mines such as the nearby Inkerman and Black Prince collieries. These structures have undergone restoration work.

Throckley coke ovens:

The ruins of beehive coke ovens at Newburn Riverside Park on the northern bank of the Tyne. Coal from the nearby Throckley Isabella Colliery was used to produce the coke. Much of it went to the nearby Spencer Steelworks at Newburn. The first ovens were built at the site in 1869 and more were added in later years. Today, the remains of five can be seen in woodland off Coach Road, close to Blayney Row. The ovens ceased operation in 1924-26.

Dunston Staiths:

Staiths were coal-loading jetties or platforms found at ports and harbours of the North-East, including the rivers Tyne and Wear, the harbours of Blyth and Seaham and West Hartlepool docks.

Ships, known as "colliers", took on coal as they lay alongside the staiths, which were connected to the railways. Pictured are the surviving Dunston Staiths, ten miles up the River Tyne at Gateshead, which opened in 1893. A second set of staiths was added to the facility in 1903. They loaded their last shipment of coal in March 1980.

A section of this impressive wooden structure was damaged by fire in November 2003, and the staiths were again hit by fire in 2010 and 2019. However, what survives is a monument to the once booming coal trade of the Tyne. Dunston Staiths is believed to be the largest wooden structure in Europe.

Main Dyke Stone, South Gosforth, Newcastle:

Pictured is a stone with an inscription which seems to mark the line of the Ninety Fathom Dyke, a geological fault which "threw down" the coal measures several hundred feet below where they otherwise would have occurred. This created a problem in the sinking of Gosforth Colliery in the early 19th Century. The pit had to be sunk to a greater depth than originally envisaged to reach the level of the "thrown down" coal and a horizontal drift cut through the rock disordered by the fault. The line of the fault runs across northern districts of Tyneside.

The Main Dyke marker stone carries an inscription which today is difficult to interpret, but it may mean that the drift was cut through the rock to a distance of 349 yards from the bottom of the pit in June 1828. It seems, however, that the drift may have been extended further. The stone can be found close to Aidan Walk, South Gosforth, although it is likely that it has been moved and is not in its original position.

Gosforth Colliery, which opened in 1829 after the completion of the drift, was situated at South Gosforth, a little to the north of Haddricksmill Bridge and near the western bank of the Ouseburn.

The 90 Fathom Dyke also cut across the workings of Backworth Colliery, near Shiremoor. Two of the colliery's pits, the Eccles and the Maud, were separated by the fault. One of the pits was much deeper than its neighbour because of the geological fault. However, the shafts of the two pits were barely 100 yards apart. The colliery also featured a third shaft.

Neville Hall, Newcastle:

Neville Hall, pictured opposite on the left, and the statue of steam locomotive pioneer George Stephenson, situated close to Newcastle's Central Station. Neville Hall is the base of the North of England Institute of Mining and Mechanical Engineers, which was established in 1852 to promote safety in the pits and to set up an educational library on mining and mechanical engineering.

The institute today plays an important role in preserving the heritage of the collieries in the North East and provides an important source of information on many aspects of mining, engineering and geology.

Neville Hall, a superb example of a Gothic Revival building, was completed in 1872 to the designs of architect A.M. Dunn. It is a Grade II* listed building and contains the Nicholas Wood Memorial Library which features an important collection of papers, maps, journals and books relating to the subject.

Nicholas Wood was a leading 19th Century colliery viewer (manager) and was the first president of the institute. He was an acknowledged expert on mining engineering, geology and safety and was a close colleague of George Stephenson at Killingworth Colliery. A fine statue of Wood occupies a prominent position in the library.

The statue of George Stephenson, opposite the Royal Station Hotel and close to Neville Hall, is the work of John Graham Lough (1798-1876). It stands at the junction of Neville Street and Westgate Road.

Stephenson, born at Wylam on the banks of the Tyne, is generally acknowledged as the leading pioneer of the steam locomotive and railways. Although he was not the inventor of the first locomotive, he honed the invention into workable and practical form. His first "travelling engine" was completed at Killingworth Colliery workshops in 1814. He was enginewright at the colliery, but also did work for other pits. One of Stephenson's early triumphs was to engineer the Hetton Colliery railway, which was linked to staiths on the River Wear. His locomotives were used on part of the line.

George, his son Robert, and several business partners founded the first steam locomotive-building factory in the world, the Forth Street Works, Newcastle, in 1823-24. It was at this works that the famous locomotive Rocket was built for the Liverpool and Manchester Railway. It also supplied locomotives for the pioneering Stockton to Darlington Railway.

The pitmen's trade unions

Red Hill, Durham:

Pictured (right) is Red Hill, the impressive Durham Miners' Association (DMA) headquarters in Durham City. It is also known as the Miners' Hall.

Early meetings of the DMA, the pitmen's trade union, were held at the Market Tavern in the city's Market Square. Its first purpose-built headquarters was opened in North Road, Durham City, in 1875. The building still stands, although it has been put to other uses. Its façade displayed statues of union pioneers, including William Crawford, the DMA's first leader, who became its president and general secretary.

Also depicted in stone were Alexander Macdonald, the Scottish and national miners' union leader and MP for Stafford, W.H. Patterson, who succeeded Crawford as DMA leader, and John Forman, who was an agent of the union, the name given to its most senior full-time officials, and who served as president. Forman earned great respect by taking part in mine rescue work following disasters, including the Seaham disaster of 1880 in which 164 miners lost their lives.

However, the North Road building proved too small for large meetings of delegates from the lodges throughout County Durham and in the early 1900s it was decided that a new headquarters was needed. A site at Red Hill in Durham City was chosen and the DMA commissioned North East architect Henry Thomas Gradon to design a new miners' hall in which union delegates could assemble together and which would contain committee rooms and offices for staff. Few could have been disappointed with the result.

Gradon created a fine building with a red and buff stone façade, arched lower windows and a green dome surmounting

the building's profile. Built from the contributions of the pitmen, it was completed in 1915. The county's mines employed around 165,000 in 1913.

Today, County Durham is known as "Land of the Prince Bishops", yet it might just as appropriately be called "Land of the Pitmen".

The tall, imposing statues from the old North Road building flank one side of the driveway at Red Hill (pictured below left). The garden in which they stand is a memorial to the miners who died in the Easington and Eppleton disasters of 1951.

Red Hill Council Chamber:

Red Hill, or the Miners' Hall, features a large chapel-like Council Chamber (pictured right) for delegate meetings and conferences. The chamber is lined with oak panelling and could accommodate a large number of seated delegates in curved rows of oak, pew-like seats. The overall effect is similar to that of a large Methodist chapel and this is hardly surprising considering the influence of Primitive Methodism among the early miners' leaders and officials. In this large room the council of the DMA met, which was formed from representatives of all the lodges. By the early 1900s there were almost 200 collieries in County Durham. Appropriately, Red Hill is sometimes referred to as the 'Pitman's Parliament'. The seats in the Council Chamber have recently undergone restoration work as part of a scheme to expand Red Hill's new role as a centre of culture and heritage, and as a hub for the mining community.

Miner at the entrance:

The statue of a kneeling miner adorns a gate pillar at Red Hill. This is one of two sculptures of pitmen which flank the entrance to the DMA headquarters, also known as the 'Pitman's Parliament'. They were created by well-known North East artist and former Whitburn Colliery miner Bob Olley.

Union pioneer:

A statue of William Crawford which flanks the driveway of the Miners' Hall, Red Hill. William Crawford was the DMA's first general secretary and worked tirelessly to build up the strength of the union. His statue is one of four which flank the driveway. Crawford was elected "Liberal-Labour" MP for Mid Durham in 1885. He died in 1890. All four statues were sculpted by J. Whitehead.

The Putter:

Pictured is the sculpture of a miner - a putter - lifting a derailed coal tub back on to the lines. It can be seen in the memorial garden at Red Hill. Entitled The Putter, this powerful work was created by Brian Brown, an artist and former Silksworth pitman.

Courageous veteran:

The memorial to Tommy Ramsay in Blaydon Cemetery. He was a veteran of the North-East miners' strike of 1844. This dispute led to mass evictions of miners and their families from their tied housing. Tommy Ramsay was an ardent trade unionist and a great recruiter for the Durham Miners' Association in its early days.

In the face of opposition from employers, he toured the mining villages with his crake (rattle), encouraging men to attend meetings and join the union. Uttering slogans like "Lads, unite and better your condition", he is remembered as a brave man. Tommy died in 1873. His memorial statue shows him in what seems to be a speaking pose, as if still stressing the importance of trade unionism. Tommy was clearly held in high regard by his fellow pitmen. A plaque on the memorial states that it was erected by the miners of Durham.

Burt Hall, Newcastle:

Pictured on the next page is Burt Hall in Northumberland Road, Newcastle, the former headquarters of the Northumberland miners' union. Opened in 1895, it was named in memory of Thomas Burt, who led the union for many years during the 19th Century.

Born in 1837 at Murton Row, a small group of cottages near North Shields, Thomas started work at the age of 10 as a trapper boy at Haswell Colliery, operating ventilation doors. He later worked as a pitman at several other mines, including Sherburn and Seaton Delaval collieries.

After becoming an active trade unionist, Burt was elected secretary of the Northumberland Miners' Association in 1865 at the age of 27. At the 1874 general election he stood as a radical labour candidate for Morpeth, with Liberal Party support. Burt won and became one of the first two miners elected to Parliament.

In 1892, Liberal Prime Minister William Gladstone appointed him Parliamentary Secretary to the Board of Trade and he served in this post for three years.

Much respected for his wit, kindness and negotiating skill, he also became Father of the House of Commons. A stone plaque on the façade of Burt Hall pays tribute to Thomas Burt's dedicated service to the Northumberland miners. He died in 1922 and is buried in a family grave in Jesmond Old Cemetery, Newcastle.

Another plaque on the facade commemorates William Straker, who became the leader of the Northumberland miners' union after the retirement of Burt from the post in 1913. He had started work as a teenager at Widdrington Colliery and later became a hewer at Pegswood Colliery.

Straker actively promoted better education, health and housing facilities in the North East. He also played a leading role in the Northumberland Aged Mineworkers' Homes Association

and served on the House Committee of the Royal Victoria Infirmary, Newcastle.

A strong believer in negotiation rather than strikes, he was also a Christian socialist and pacifist who spoke out against the First World War. William Straker was made a CBE in 1930. His portrait occupies a central place on the Pegswood miners' banner.

Burt Hall is now a part of Northumbria University. The building is surmounted by the fine statue of a miner with a pick over his shoulder (pictured). The work is based on one of the figures in the popular painting by Ralph Hedley of two pitmen entitled Going Home. The pithead of Blaydon Main Colliery is shown in the background of the painting.

The inner main entrance door of the building is decorated with stained glass windows depicting two safety lamps, and on the upper floor a meeting hall contains a fine stained glass window featuring shields with images of picks, shovels, a lamp, tub and other mining emblems.

The origins of the Northumberland miners' union can be traced to Christmas Day 1862 when a mass meeting of pitmen gathered at Horton. In January 1863 delegates of the miners met and formed a Northumberland miners' association. Later that year pitmen from several County Durham collieries joined the new union to form a joint association of the two counties. However, in 1864 the Northumberland men seceded from their Durham counterparts and founded the Northumberland Miners' Mutual Confident Association.

Thomas Hepburn gravestone:

Pictured, bottom right, is the gravestone of pioneer North East miners' trade union leader Thomas Hepburn, situated in St Mary's Churchyard, Heworth, Gateshead. A man of peaceful intent, Hepburn led the Northumberland and Durham pitmen's strikes of 1831 and 1832.

The first strike secured a reduction in the long hours which boys worked underground and the abolition of tommy shops. These shops were owned or controlled by the coal owners and the pitmen were forced to buy provisions from them.

A great advocate of improved education for miners and their children, Thomas Hepburn believed that every pit community should be provided with a library.

Following the 1832 strike, which ended with defeat for the union, the pioneer leader was reduced to destitution and eventually forced to give up union activity in order to obtain a job - at Felling Colliery. He died in 1864 and his portrait adorns several Durham miners' banners. A Primitive Methodist lay preacher, Hepburn had worked as a miner at Urpeth, Fatfield, Jarrow and Hetton collieries.

Each year, North East miners pay tribute to this trail-blazing leader at an annual service at St Mary's Church at Heworth, Gateshead. Pitmen's banners are on display inside the church for the occasion and afterwards they are taken to Hepburn's graveside where Gresford, the miners' anthem, is played by a brass band and wreaths are laid. Gresford was composed by Tyneside miner Robert Saint in 1934. He worked at Hebburn Colliery. Our other photograph shows the first wreath being laid at the grave in 2017.

Hepburn's gravestone carries the message: "Shorter hours and better education for miners." A further inscription states that he led the 1832 strike with "great forbearance and ability". The stone was erected by "the miners of Northumberland and Durham and other friends".

Faithful in all things:

Our picture shows the inscription on the grave of Peter Lee and his wife, Alice, in Wheatley Hill Cemetery. Peter was born at Fivehouses, Trimdon Grange,

IN LOVING MEMORY OF
PETER
THE BELOVED HUSBAND OF
ALICE LEE
WHO DIED JUNE 16TH 1935, AGED 70 YEARS.
ALSO OF THE ABOVE ALICE LEE
WHO DIED 17TH OCT. 1944.
AGED 80 YEARS.

and was a miner at several pits in County Durham. He served as checkweighman at Wheatley Hill and Wingate collieries.

Peter Lee went on to start a distinguished career in local government, becoming chairman of Wheatley Hill Parish Council. Later, he was elected to Easington Rural District Council. In 1909, Lee was elected to Durham County Council and in 1919 became its chairman. It was the first Labour controlled county council in Britain.

The former pitman became general secretary of the Durham Miners' Association in 1930, holding this position until his death in 1935. The New Town of Peterlee was named in his honor.

Lee and his wife lived at Wheatley Hill for 19 years and his efforts brought link roads, drains, a sewerage system, street lighting, an isolation hospital and a cemetery to the Wheatley Hill area.

He also played a leading role in pressing for the creation of the Burnhope Reservoir in Upper Weardale to improve water supply to County Durham. He is buried in the village cemetery he had campaigned to provide. An inscription on the grave reads: "Faithful in all things."

Strike and eviction:

A plinth surmounted by a coal tub on rails at Sunnybrow, Willington, County Durham, commemorates the "Rocking Strike" of 1863. It is situated on the edge of the village on a green expanse of hillside overlooking the River Wear. The dispute was over payment for tubs of coal and involved the collieries of Sunnybrow, Oakenshaw and Brancepeth.

At Sunnybrow, the pitmen found the tubs they filled to capacity underground were reaching bank with levels of coal deemed unsatisfactory by the keeker, an official employed by the owners. Many tubs were confiscated - "laid out" - because of this. The keeker was only too happy to have the tubs laid out because he received commission. The miners complained that the tubs were not large enough to hold the amount of coal they were expected to put in them.

According to miners' historian Richard Fynes (The Miners of Northumberland and Durham: A History of Their Social and Political Progress, 1873) the coal was packed as closely as possible, but in the low places in which men had to work the tubs were jolted as they went outbye, the coal being shaken down. To combat this effect the men began the practice of "rocking" the tubs so that the coal slipped down and the tubs could be topped up before being taken to bank. This rocking process was extremely hard and "painful" work.

Alternatively, a man might strike the tub with his mel (long-handled, heavy hammer) to take the coal down before being topped up.

The pitmen went on strike over the issue, calling for wages to be paid according to the weight of the tubs of coal rather than on the basis of the number of "full" tubs. The dispute led to evictions of miners and their families from their colliery houses.

As well as the strike, the memorial also commemorates all the men and boys who lost their lives in the mines of Sunnybrow, Oakenshaw and Brancepeth. It was unveiled by NUM president Joe Gormley in 1976.

A grandson remembers:

Jack Fletcher, grandson of the secretary of the Chopwell miners during the Great Lockout and General Strike of 1926, is pictured with the Chopwell Colliery pit wheel memorial. Jack's grandfather was Harry Bolton, who served a short jail sentence for his militancy during the dispute but later became chairman of Durham County Council.

During the Great Lockout Britain's miners were fighting against reductions in wages and longer working hours. The wheel, situated in the centre of the village, commemorates all those who worked at the Chopwell mine during its lifetime (1896-1966). Next to it is a mining heritage stone with an inscription from a Walt Whitman poem: "We take up the task eternal, the burden and the lesson. Pioneers! Oh pioneers!" These words are also inscribed on the Chopwell miners' banner. They are, of course, very appropriate for trade unionists.

The banner carries the portraits of James Keir Hardie, Karl Marx and Lenin. The faces of Marx and Lenin were first painted on the standard around 1924 when Chopwell miners felt the Russian Revolution offered hope of a better world for ordinary people and could not have envisaged the emergence of Stalin as a ruthless dictator. The portrait of Keir Hardie, a democratic socialist, is accorded central place on the standard.

Jack Fletcher, a school teacher all his working life, writes of Harry Bolton: "The people of Chopwell knew my grandfather as a very busy 'public man', ever ready to help them with their problems, be they mundane or complex. The miners saw him as a fighter for their cause, against the mine owners, both at government level, as when he gave evidence to the Sankey Commission of 1920, and regional level at meetings of the DMA and Durham or Blaydon council.

"He was the leader, alongside Will Lawther, of the famous Chopwell lockout. A well-read, literate man, during the 1926 dispute he was the author of an illicit broadsheet, The Northern Light, the tribune of the strikers. He edited The Northern Light just as Winston Churchill oversaw production of the British Gazette."

Jack points out that in private life Harry Bolton was a kindly, loving father of five children and grandfather of six. A staunch Methodist (he called his son Wesley) his life revolved around the chapel, where he was superintendent."

Because of the militancy of its miners during the 1926 dispute Chopwell was dubbed "Little Moscow" by some, but Jack strongly rejects this nickname as inaccurate. He stresses that the driving force behind the resistance of the Chopwell men was Keir Hardie's Independent Labour Party.

Miners' welfares:

Miners' welfare centres, also sometimes known as institutes, are found in many communities throughout the Great Northern Coalfield. Although every deep mine in the North-East has closed, a considerable number of these buildings or "welfares" as they are called, are still standing and now serve as general community centres.

The 1920 Mining Act established the national Miners' Welfare Fund, a recommendation of the Sankey Commission. Colliery owners contributed a levy of one penny per ton of coal produced towards this fund, which financed schemes to improve the life of the pitman and his family. From 1926, a levy was also placed on coal royalties.

A Miners' Welfare Commission was set up to administer the fund and money was allocated to each coalfield area to help provide welfare buildings, sports fields, children's playgrounds, pithead baths, canteens and other facilities such as libraries. Locally-based district committees, with members from both unions and management, decided on how their allocation of money was spent.

Welfares were among the most important schemes to result from this initiative. They became social centres for many events, including dances, shows, wedding celebrations and games such as billiards, snooker and table tennis. They were also venues for union meetings. Frequently a reading room was provided in which newspapers and other publications were made available for the miners. Some welfares featured a library.

Perhaps the most impressive surviving welfare is the one at Easington Colliery, which is still a thriving social centre. This large building, opened in 1934, contains a fine dance floor and stage. The former miners' reading room has been converted into a lounge and is used for meetings. The former ladies' cloakroom has been transformed into a "youth room" with IT and music facilities. A large billiards and snooker room is another feature of the centre.

Easington Colliery, a very productive mine which extended several miles out under the North Sea, closed in 1993. Scenes for the highly successful film Billy Elliot were shot in the village.

Easington Colliery Miners' Welfare

Coastal institute:

Cambois Miners' Welfare Institute opened in 1929. Immediately behind this building is the village's social club, opened in 1910, and next to the club are Aged Miners' Homes. The welfare includes a hall on the upper floor which was used as a cinema. On the ground floor are rooms which were used for billiards, snooker and union meetings as well as a reading room where newspapers were available for the miners. Many of the pit housing rows at Cambois have been demolished.

Cambois Miners' Welfare

Banners - tradition and heritage

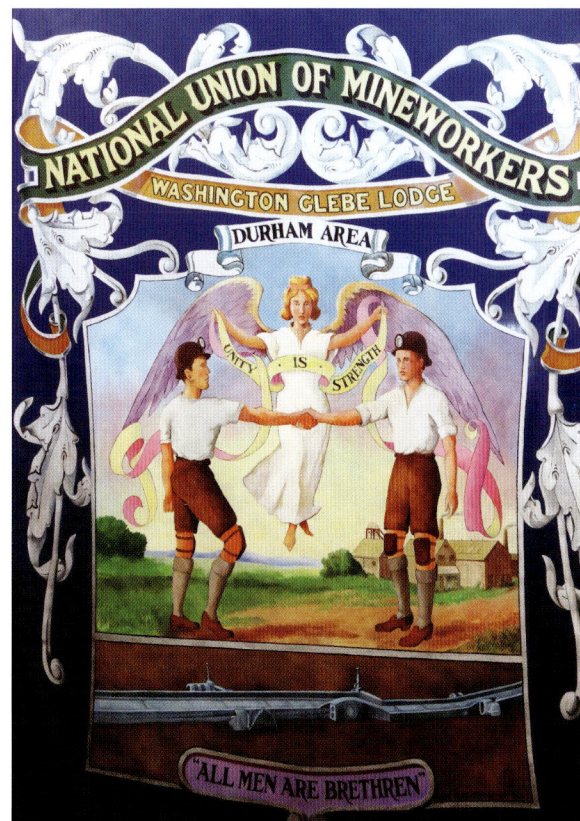

Miners' union banners are an extremely important part of the heritage of the pit communities of Northumberland and County Durham. Emblazoned across their fabric are themes and mottoes which reflect the history of the pitmen and their values.

These colourful standards, painted on both sides, are rich in meaning and significance. Many proclaim the importance of trade unionism and bear mottoes of solidarity. The favourite is "Unity is strength." A number of other mottoes bear messages of peace, brotherhood and help for those in need.

The craftsmanship and artistry involved in producing these impressive icons of trade unionism should never be underestimated. Banner makers over the years include such well-known names such as George Tutill of London and later of Chesham, Chippenham Designs of Norfolk, Bearpark Artists' Co-operative, Durham Bannermakers of Durham City, Aidan Doyle, of Great Northern Banners, Newburn, and Turtle and Pearce of London,

George Tutill can trace its origins back to 1837 and for many years in the 19th and 20th centuries produced more banners for the North-East miners than any other firm. The business is still in existence and is part of the Flagmakers Group.

Banners were made according to designs submitted by the miners' union lodges or selected from themes in the Tutill catalogue. These catalogue themes have often been replicated on successive banners down the years.

Perhaps the most impressive Tutill banners, and to many people the most attractive, were those made from huge silk squares or rectangles

All men are brethren:
A Washington Glebe lodge banner bears the figure of an angel and two miners shaking hands in friendship. The two mottoes are "All men are brethren" and "Unity is strength". The standard was made by Bearpark Artists' Co-operative and is mainly based on an earlier theme on a Tutill banner. A "Panzer" conveyor belt used in the pits to carry coal away from the face, is illustrated in the lower section of the standard.

which were woven on a large Jacquard loom at the firm's studios. These very advanced looms used a punch-card method to weave into the silk an elaborate design which surrounded a central, blank roundel or shield-like panel on which the main picture or pictures would be painted. Fabric like this, with a woven-in pattern, is known as damask.

The woven surround, which could be of various colours, featured an ornate design of foliage with curving leaves, plant shoots and horns of plenty with fruit.

Following weaving the picture or pictures illustrating the main theme of the banner would be painted into the blank central space. Spaces were also left for the painting of the lettering, including the motto or message.

Great care was taken before the designs were painted on to the fabric. Special preparations were needed to make the final surface more durable. It was important that the paint did not crack, peel or come away easily.

Tutill employed teams of artists. Some might have expertise in figures or portraits, others in landscapes and buildings. There would also be painters who concentrated on the lettering.

The damask silk banners woven on the Jacquard loom were the most expensive provided by Tutill. However, cheaper banners were available for miners' union lodges with a smaller membership and therefore less money. These did not feature the woven-in leafy plants design. Instead, the entire design, including the surround, would be painted on to a plain silk square. The painted surrounds differed to some extent from the woven-in pattern but usually featured leafy, foliage-like ornamental designs and often cords with tassels or ribbons were depicted.

Unfortunately, the Tutill studios in London were bombed during the Second World War. The damask silks were never produced again.

Despite this blow to a long tradition, a few of these beautiful woven silks survived the bombing raid and former County Durham miner Dave Gibson stepped in to help save some of them for posterity.

In recent years they have been used as the basis for several new miners' banners. The damask pattern is instantly recognisable on these "new" standards, which feature recently painted illustrations in the central roundels or panels.

However, today most banners are not made from silk. Instead, artificial fibres, such as polycotton, are used. In the vast majority of cases the designs are painted on, but they still frequently display traditional motifs and patterns which go right back to Tutill and other early banner makers. Yet despite this tendency to adhere to traditional themes, new motifs are sometimes selected for one or both sides.

New banners are made to replace old, worn ones which have become too fragile to parade. The cost of making these standards generally runs into thousands of pounds and the Heritage Lottery Fund frequently helps with finance. Funding is also provided by the former pit communities themselves. Banner groups are formed within the communities to raise the money and apply for funds. In addition, a considerable number of older banners have undergone restoration work.

However, whether the banners are old or new they are greatly cherished and well cared for by former mineworkers, their families, friends and the Durham Miners' Association. Some of the older County Durham examples are now preserved at Beamish museum and others are kept at Red Hill, the headquarters of the Durham Miners' Association in Durham City. Many others are scattered throughout the coalfield and have found homes in miners' welfares, community centres, churches, council offices and schools.

It is recognised that these icons of trade unionism represent each pit community as well as the lodges. They symbolise the brotherhood of the miners, their essentially humanitarian values and the warm spirit of those communities.

Yet the history of the banners is not confined to symbolism. They have played very important roles in real life, helping to recruit men to the union, acting as inspirational rallying points, strengthening the miners' unity during times of strike and

Elemore banner at gala:
The Elemore banner, with band members in the foreground, at the Durham Miners' Gala in 1956. They are waiting to parade on to the old racecourse at Durham. The banner is draped in black, indicating at least one fatality at Elemore Colliery during the preceding year. (Photo by Derrik Scott)

Picnic scene at Bedlington:
Crowds and banners at the Northumberland Miners' Picnic at Attlee Park, Bedlington, during the late 1950s or early 1960s. Bedlington was the scene of the Picnic for many years. (Photo by Derrik Scott)

lockout, giving solace in times of ordeal and paying due tribute to the departed at funerals.

The standards have frequently carried poignant reminders of the dangers faced by pitmen. At the annual Durham Miners' Gala, black crepe was draped across the top of a banner if one or more miners from that union lodge had been killed in a pit accident during the preceding year.

Black crepe was also draped across standards at funerals. This solemn rite is still continued today if leaders, officials or other stalwarts of the union have passed away.

The greatest showcase for the pitmen's banners is the annual Durham Miners' Gala - or Big Meeting as it is affectionately known - which has been held in most years since 1871. The popular Gala parade through the streets of Durham City still takes place despite the closure of every deep mine in the Northumberland and Durham coalfield.

The Big Meeting is organised by the Durham Miners' Association (DMA) in partnership with the Friends of the Durham Miners' Gala (known as the Marras). The DMA was founded in 1869 as the pitmen's trade union.

The Gala is a remarkable survival and still attracts many thousands of people. For most of its history it has been held on the second Saturday in July.

The Gala begins with former miners, their families, friends and well-wishers assembling at various points in Durham City. The many lodge contingents, with their banners held proudly aloft, then march through the streets of the city, parading down the historic thoroughfare known as Old Elvet. They are accompanied by brass and silver bands playing a medley of stirring tunes. Bands pause at the County Hotel to serenade the Gala's leading guests and DMA leaders who stand on the hotel's balcony. The atmosphere is always good humoured.

The contingents with their banners and bands continue down Old Elvet to the former racecourse by the River Wear,

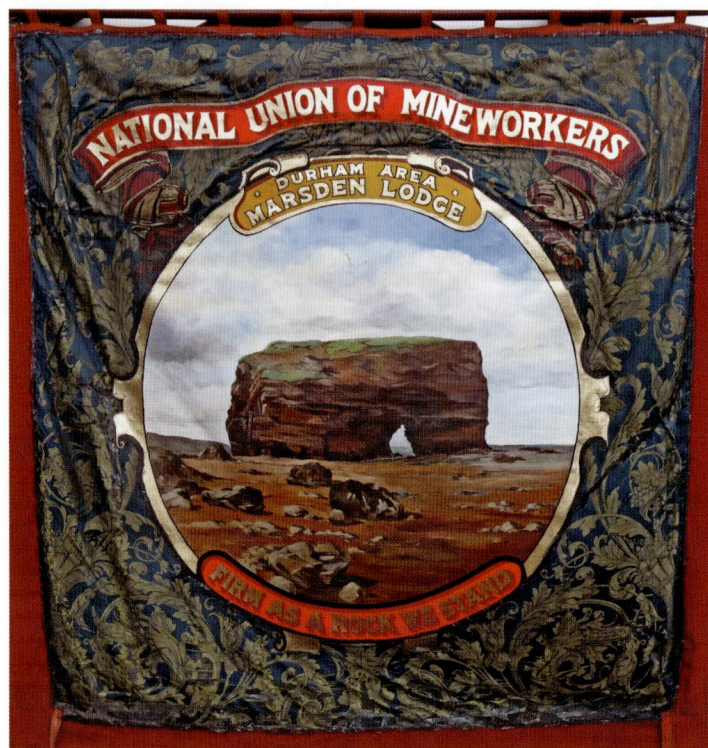

Firm as a rock we stand:
The Marsden Lodge banner, showing an impressive view of the Marsden Rock, a well-known sea stack. 'Firm as a rock we stand,' reads the motto. This veteran banner dates to the early 1950s and was made by the famous banner-making firm of George Tutill.

Marsden Colliery, also known as Whitburn Colliery, closed in 1968. The mine extended out under the North Sea. The mining village of Marsden, much of it situated next to the northern side of Souter Lighthouse, has now vanished. The colliery was sited to the south of the lighthouse.

Restoration work on the banner was carried out by Durham coalfield artist Bob Ord and retired mineworker Billy Middleton, of Thornley, County Durham. They have restored a considerable number of other veteran standards.

where there are stalls, tents and a funfair. The banners are tied to the fences surrounding the field, creating a colourful display.

Before the speeches, the Mayor of Durham, accompanied by a ceremonial bodyguard, welcomes the men, women and children to the city. Robert Saint's famed miners' anthem, Gresford, is played in front of the platform by a chosen band.

The speeches on the racecourse have always been a central feature of the day. Many renowned Labour politicians and trade union leaders have addressed the Gala crowds.

Later, the banner contingents and bands begin the march out from the city centre. They again parade down Old Elvet and over Elvet Bridge, still serenading the onlookers.

A considerable number of the bands which play at the Big Meeting can trace their origins to the collieries and many still retain the name 'colliery' in their titles. The music of the bands is an essential element of the day. The proud tradition of musicianship in the pit communities is a strong one and today there are many talented young people who play in the bands, helping to continue that tradition.

The first Gala was held at Wharton Park, Durham, on August 12, 1871. It was essentially a mass demonstration of the miners' solidarity, intended to send a message to the employers that the men were united. About 4,000 to 5,000 pitmen and their families attended. Every Gala since then has been held at Durham's old racecourse by the River Wear.

As well as a miners' trade union rally, the Big Meeting also developed into a major family day out, an occasion for fun and enjoyment, and an opportunity for people from the many scattered pit communities of County Durham to meet and socialise.

A record attendance of over 300,000 was reported in 1951. In some other years during the 1950s estimated attendances were around 200,000.

Today, the event still attracts many thousands, and has gone from strength to strength in recent years. As well as the National Union of Mineworkers, members of many other trade unions take part, often bearing their own banners.

Most of the Northumberland miners' banners are now preserved at the Woodhorn Colliery museum on the outskirts of Ashington. The Northumberland standards were paraded at the Northumberland Miners' Picnic, which for many years after the Second World War took place at

Bedlington. However, Morpeth, Blyth Links and Newcastle's Town Moor were among earlier venues chosen. At Bedlington, the speeches were delivered from the bandstand in Attlee Park.

The first Picnic was held in 1864 at Blyth Links, although the first to take place after the election of Thomas Burt as secretary of the Northumberland miners' union was held in 1866 at Polly's Folly, a field between Bog Houses and Shankhouse.

The Picnic featured a keenly fought brass band contest, the top trophy being the Burt Challenge Cup. To win this trophy was a major achievement.

Today, there is no parade or contest, but the Picnic now takes the form of various events at Woodhorn Colliery museum, with brass bands and other entertainment included. In addition, there is a memorial service for Northumberland miners, and a lecture organised by Northumberland NUM.

The history of North-East miners' banners stretches back into the first half of the Nineteenth Century.

One of the first references to the use of these standards by the pitmen is contained in an account of the 1831 dispute, written by Richard Fynes in The Miners of Northumberland and Durham, published in 1873: 'On the 21st April (1831) a large meeting of miners was held in Jarrow, each colliery bearing a banner, with the name of the colliery and various mottoes.'

Emancipation of Labour:

A traditional "Emancipation of Labour" theme on a veteran New Herrington miners' banner. A female figure symbolising "Progress" leads a group of men, women and children to a better life in the "Co-operative Commonwealth". The motto is a quotation from Marx and Engels: "Workers of all lands unite. You have nothing to lose but your chains. You have a world to win."

This socialist motif has been a favourite with various lodges over many years and can be traced back to a George Tutill design believed to have been inspired by a Walter Crane drawing of the early 1900s. Walter Crane was an accomplished artist and socialist. He produced many illustrations for children's books. The other side of the standard shows a view of Conishead Priory, near Ulverston in Cumbria, which was used for many years as a convalescent home for injured or sick miners.

The banner, which is displayed in Wheatley Hill Heritage Centre at the village cemetery, dates to around 1932 and was made by Tutill. The site of New Herrington Colliery, in the Durham Coalfield, is now a large country park.

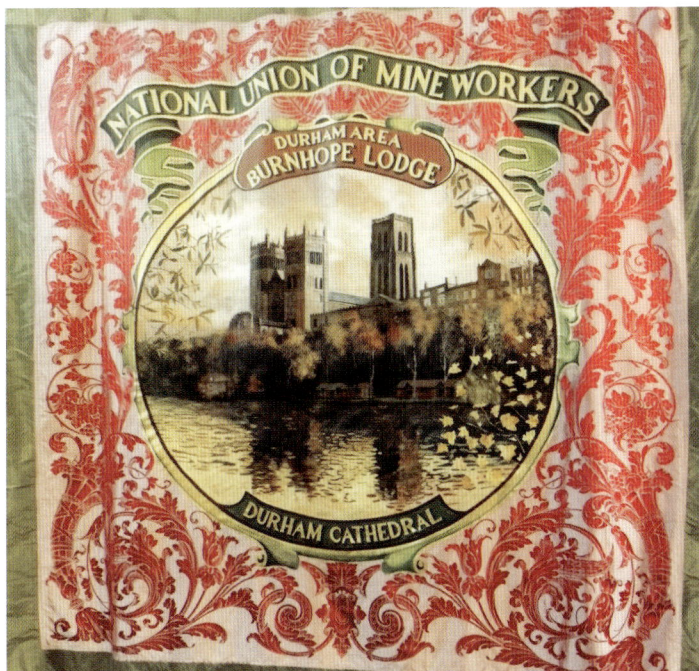

The cathedral and river:

A picture of Durham Cathedral and River Wear, with autumn colours in evidence, adorns a Burnhope lodge standard. Similar views of the cathedral and the river have been a favourite banner subject of some lodges for many years. Burnhope Colliery, near Lanchester, opened in the 1840s and closed in 1949. This banner was painted by Durham Bannermakers using a sample of damask silk woven in Tutill's City Road, London, studio more than 70 years ago.

Celebrating nationalisation:

One side of this Harton and Westoe miners' banner celebrates nationalisation of the mines in 1947. A family are seen standing at the gate to a field as they greet the new dawn of common ownership.

The other side of the standard carries portraits of James Keir Hardie, first leader of the Labour Party, A.J. Cook, who led Britain's miners during the General Strike and Great Lockout of 1926, and Clement Attlee, Labour Prime Minister.

The banner was produced by Durham Bannermakers.

Labour and Peace:

A miner representing Labour holds hands with a woman representing Peace in this close-up view of the central section of a Dawdon lodge banner. Its conciliatory tone implies a desire for peaceful settlement of disputes. It can also be read as a statement against war as a solution to problems. The theme, originating with Tutill, was used by Dawdon lodge as far back as 1933. This latest banner was produced by Durham Bannermakers.

The late Dave Guy, who served as president of the Durham Miners' Association for many years, was a Dawdon Colliery miner. Dawdon was one of three collieries in the Seaham area, the others being Seaham and Vane Tempest.

Pitman at the gate:

In this close-up view of a Crook Drift lodge banner a miner is being given the key to the gate of economic emancipation by a female figure representing trade unionism. The key represents "organisation". Crook Drift was also known as the Hole in the Wall Colliery. The mine opened in the 1930s and closed in 1964. The banner was made by Bearpark Artists' Co-operative and replicates a traditional theme. The members of the co-operative are Barrie Ormsby, Romey Chaffer and John Foker.

For all who suffered:

A grieving widow and daughter are shown by a gravestone on the East Hetton (Kelloe) lodge banner with an angel holding a wreath in the background. The inscription on the gravestone remembers all miners killed and injured at East Hetton Colliery during its lifetime. In the background, right, is the Kelloe memorial to miners killed in the Trimdon Grange Colliery disaster in 1882. Seventy four men and boys died as the result of the Trimdon Grange explosion.

The banner also carries the words: "For all who suffered." This side of the standard is based on the theme of a grieving widow by a graveside on a Kelloe lodge banner of 1873, made by Tutill. The latest banner was made by Chippenham Designs.

Education is our future:

"Education is our future" declares the words on an Easington Colliery lodge banner, accompanied by a painting of children in a school playground. The DMA has always stressed the importance of education. The Easington Colliery community garden is pictured on the other side. A pennant bearing the Yugoslav colours is always draped over the top of the banner during parades and other events. This has been carried on the banner ever since a Yugoslav diplomat attending the funeral of miners who died in the Easington Colliery disaster of 1951 placed a pennant bearing his country's colours at the gravesides. This standard was produced by Durham Bannermakers.

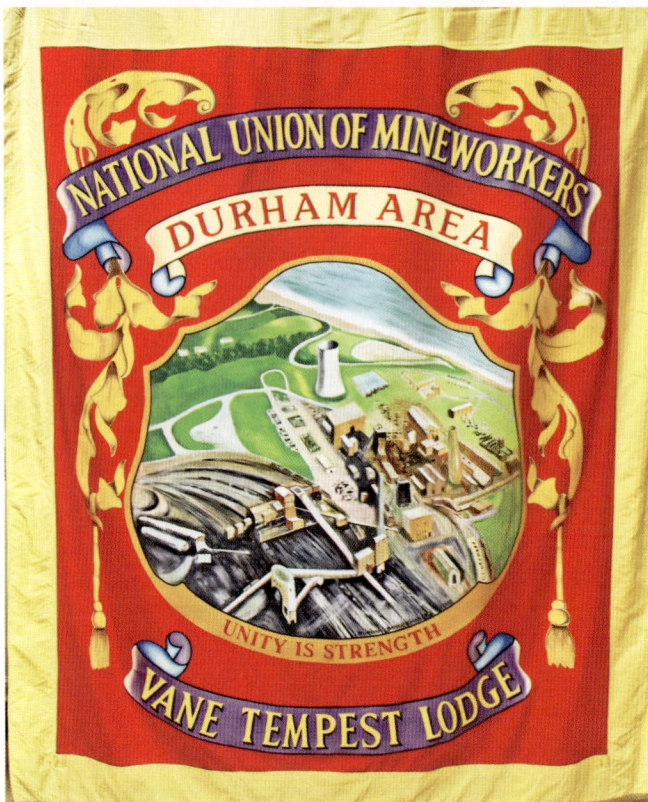

Unity is strength:

The Vane Tempest lodge banner carries a fine aerial view of the colliery, with the sea visible to the top, right. It is accompanied by one of the most enduring mottoes, "Unity is Strength." Vane Tempest Colliery closed in 1992, bringing an end to mining in the Seaham area. The banner was produced by Durham Bannermakers and replicates an earlier theme on a banner by Turtle & Pearce.

Pictured below is a sculpture by Michael Johnson based on the pithead skyline of Vane Tempest Colliery. It is located on the seafront at Seaham opposite the site of the mine. Behind the sculpture a metal inlay in the concreted ground represents the underground roadways of the 'C' seam of Vane Tempest.

Gain the Co-operative Commonwealth:

This photograph of the Northumberland Area NUM banner shows the side which features the words "The New Vision" with a female figure carrying a torch as she leads workmen, including a miner in his "hoggers" (wearing shorts and socks), towards a group of children dancing around a maypole with the words "Gain the Co-operative Commonwealth" above them. The maypole ribbons bear the words "Beauty, Fellowship, Health, Art and Science." Also featured is the call "Workers of the World Unite."

The other side of this banner, made by Chippenham Designs, displays what is perhaps the strongest imagery of any throughout the Great Northern Coalfield. A group of miners carry a coffin in a funeral procession from a colliery. Some of the men are injured. The pit is on fire.

However, emerging from the blazing pit shaft is the giant figure of a man who is pointing a spear labelled "State Control" towards a fire-breathing dragon symbolising profit and private ownership.

The message on the banner reads: "The workers' industrial union and his political Labour Party will destroy this monster." It is a call for common ownership of the mines, indicating that health and safety should be put before profit.

The banner is based on a 1920s standard made by Tutill and carried by pitmen of the Northumberland Miners' Association's Ashington Group of Collieries. The five mines in this group were Ashington, Lynemouth, Linton, Ellington and Woodhorn.

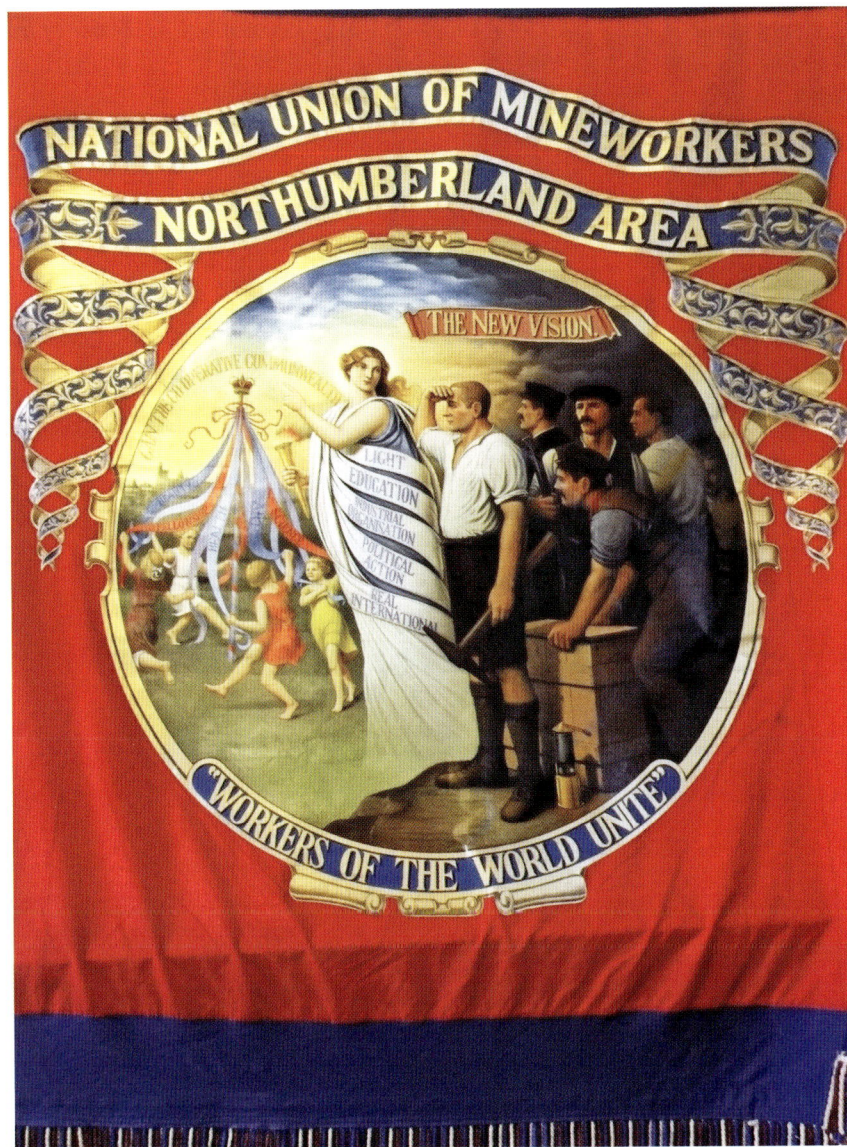

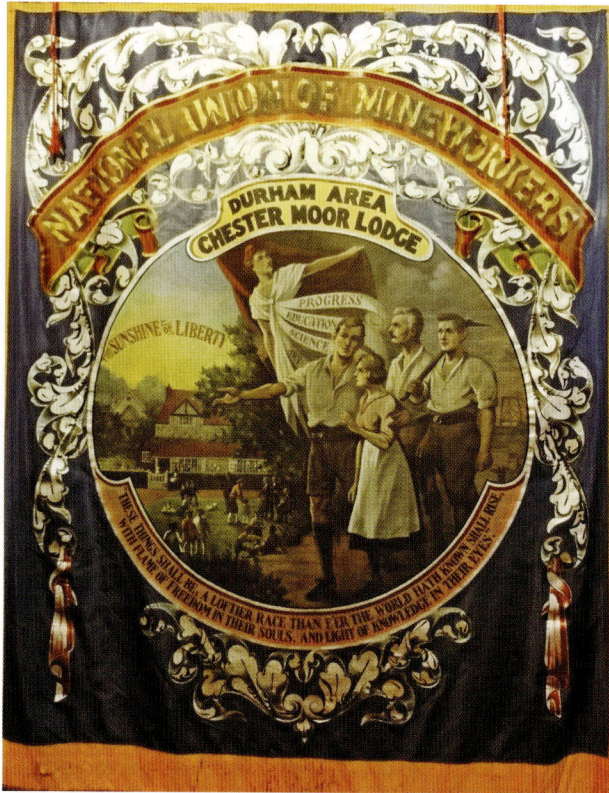

Flame of freedom in their souls:

A traditional Sunshine of Liberty design on a Chester Moor lodge banner depicts a female figure drawing back the curtain of oppression to reveal to three miners and a housewife a better life under socialism, with its ethos of co-operation and its "Sunshine of Liberty", represented by pleasant housing and happy people.

The banner carries the words: "These things shall be, a loftier race than ere the world has seen shall rise, with flame of freedom in their souls and light of knowledge in their eyes." The standard is by Tutill. Chester Moor Colliery dated back to Victorian times. It closed in 1967. The Sunshine of Liberty motif is also carried on the Wardley lodge banner.

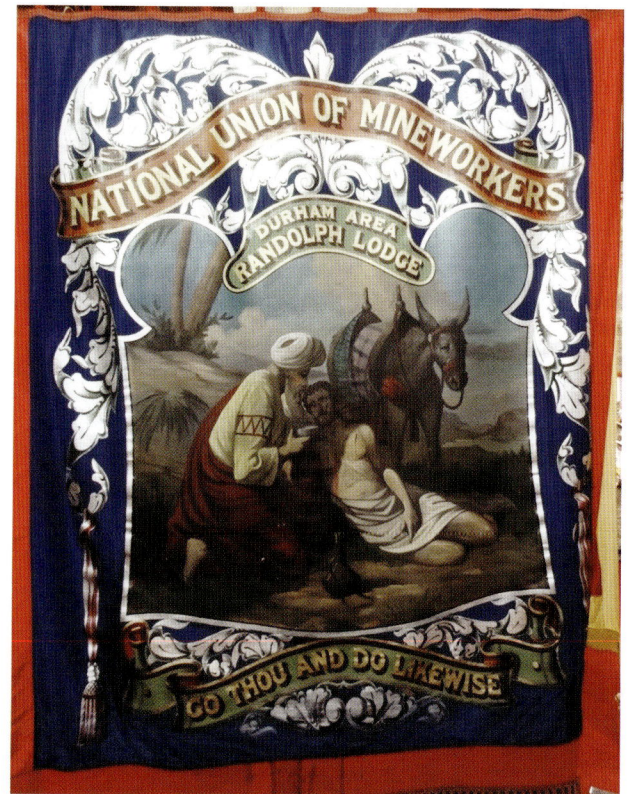

Go thou and do likewise:

The Good Samaritan theme is featured on a Randolph lodge banner with the biblical words: "Go thou and do likewise." This motif reflects the strong influence of Christianity in the pit communities. Randolph Colliery, to the south of Bishop Auckland, was situated at Evenwood, County Durham. The East Hetton (Kelloe) lodge banner also displays the Good Samaritan theme. The Randolph banner is by Tutill. A memorial, which includes images of a pit lamp and a coal gas flame, can be found at Evenwood Village. It pays tribute to those who worked in the coal and gas industries. An important cokeworks was located next to Randolph Colliery.

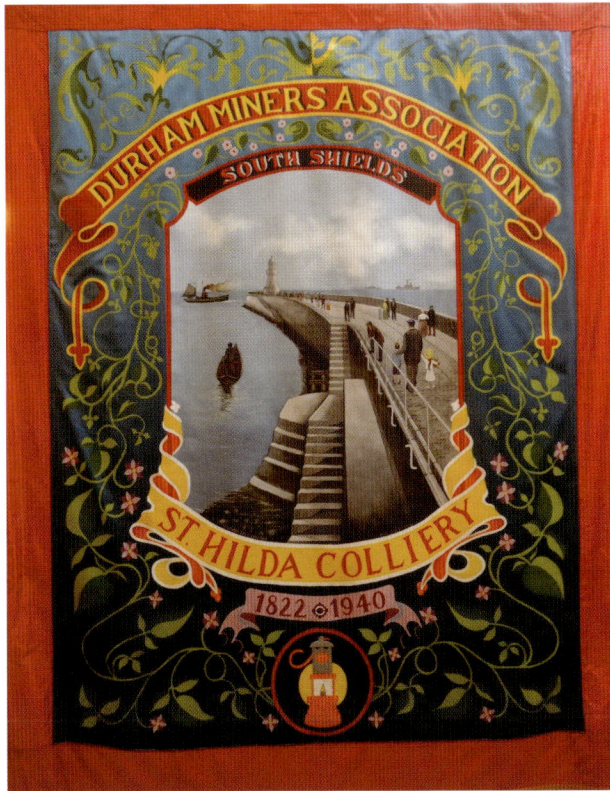

South Shields pier:

The latest St Hilda's Lodge banner of the DMA. It was produced by Durham Bannermakers in co-operation with North-East artist and former miner Bob Olley, who did the paintings for both sides. St Hilda's Colliery was situated in central South Shields and closed in 1940. This standard bears a picture of South Shields Pier at the mouth of the Tyne. It is based on a veteran St Hilda standard which also carried the same coastal image.

When the colliery closed the old banner was passed on to South Moor No. 1 Colliery Lodge, also known as Louisa Old Colliery, in the Stanley area of County Durham. The miners of Louisa Colliery, situated far from the North Sea, accordingly put the name of their lodge on the banner. The image of the pier and the sea must have puzzled some people unaware of the standard's history.

The other side of the new St Hilda's banner carries a design by Bob Olley, based on ideas suggested by children of St Bede's School, South Shields. The theme is the St Hilda's Colliery disaster of 1839 in which 51 men and boys lost their lives and the invention of the Davy safety lamp.

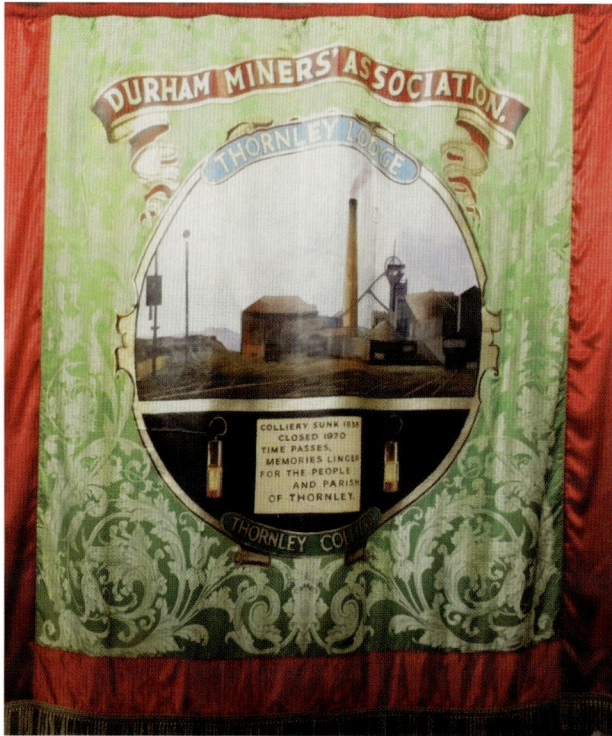

The Thornley pithead:
A painting of the pithead adorns this Thornley lodge banner. Two miners' lamps are pictured below the main illustration. The banner was painted by Durham coalfield artist Bob Ord, who comes from a mining family and lives in the village. Thornley Colliery in County Durham opened in the 1830s and closed in 1970. The banner was produced by Bob Ord and Billy Middleton and features a surviving Tutill silk.

We unite to assist one another:
The motto on this Netherton Colliery banner reads: "We unite to assist one another." Two miners are pictured flanking the globe, which is surmounted by a family. An illustration below the globe shows miners at the coalface. Netherton Colliery, near Bedlington in the Northumberland coalfield, closed in early 1974. The banner is draped in black, indicating at least one death at the colliery in the preceding year. The photograph was taken at the Northumberland Miners' Picnic in Bedlington.
(Photo by Derrik Scott)

90

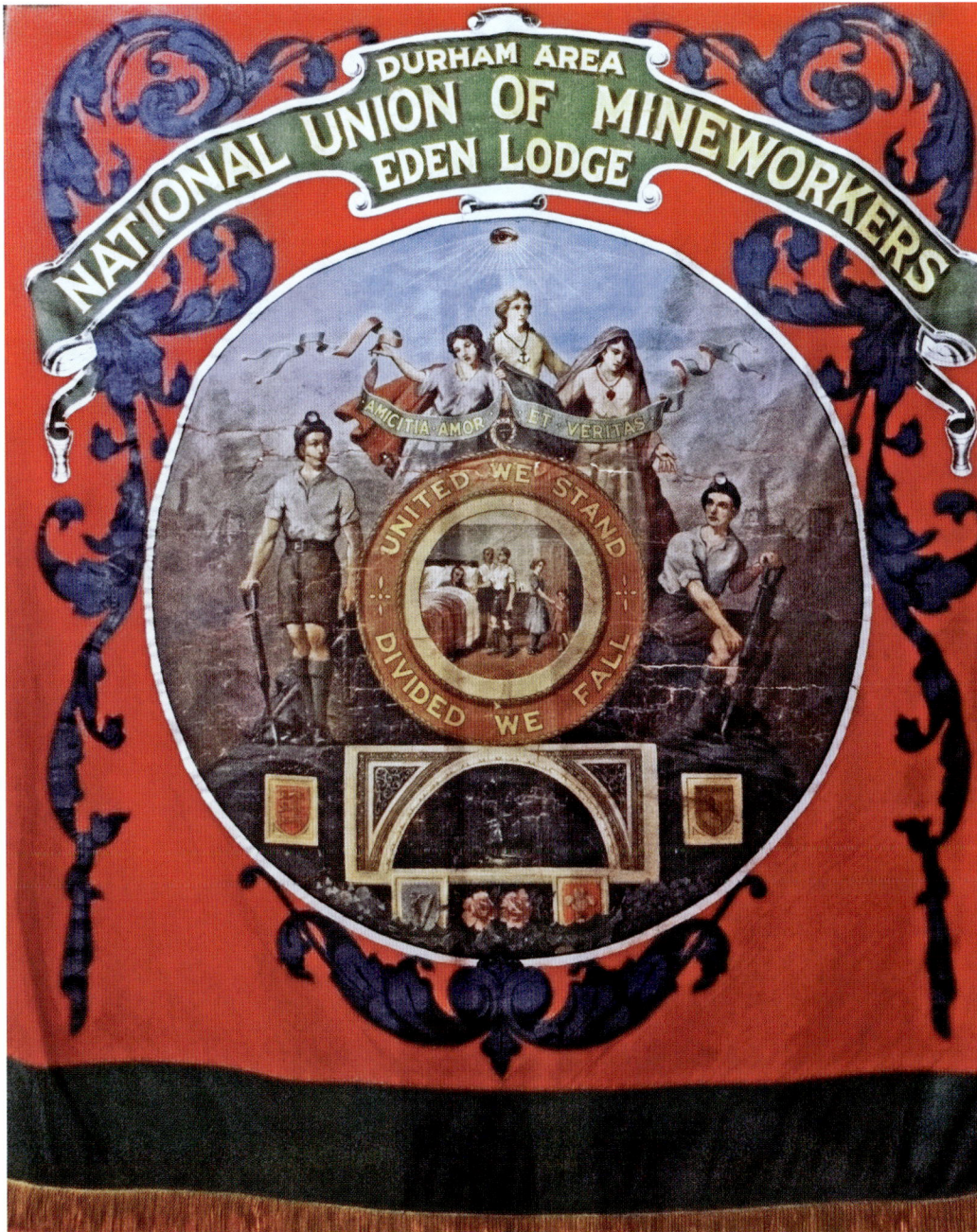

Friendship, Love and Truth:
This Eden Colliery banner, which dates to the 1950s, has many interesting details. Two miners are depicted flanking a central roundel showing another two pitmen visiting the bedside of an injured or sick comrade. The motto declares: "United we stand, divided we fall." The central roundel is surmounted by three female figures, representing Friendship, Love and Truth. In front of them is a scroll bearing the words Friendship, Love and Truth in Latin: "Amicitia, Amor, Et Veritus." The standard is by Tutill.

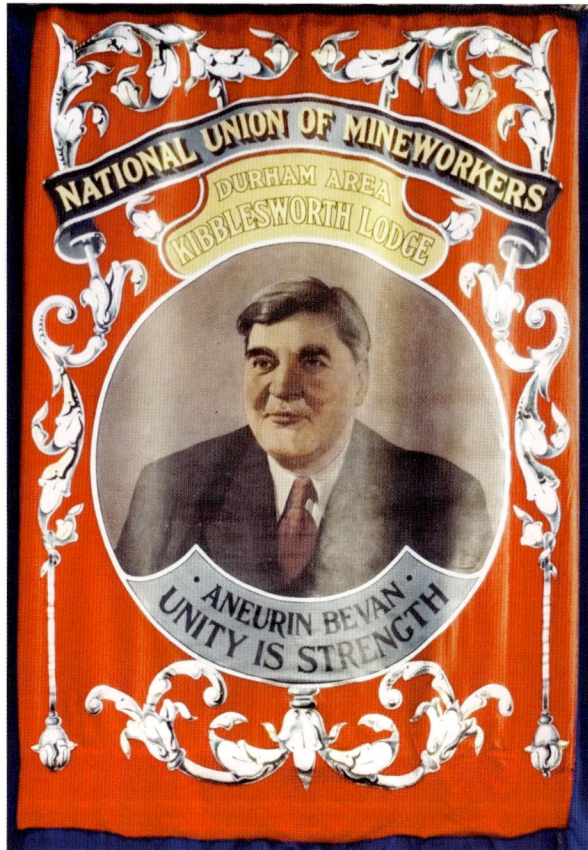

Tribute to NHS founder:
A portrait of Labour minister Aneurin Bevan, founder of the National Health Service, is pictured on a Kibblesworth Colliery lodge banner. This Tutill standard dates to 1961. "Unity is Strength" reads the motto. Kibblesworth Colliery, near Gateshead, opened in 1842 and closed in 1974.

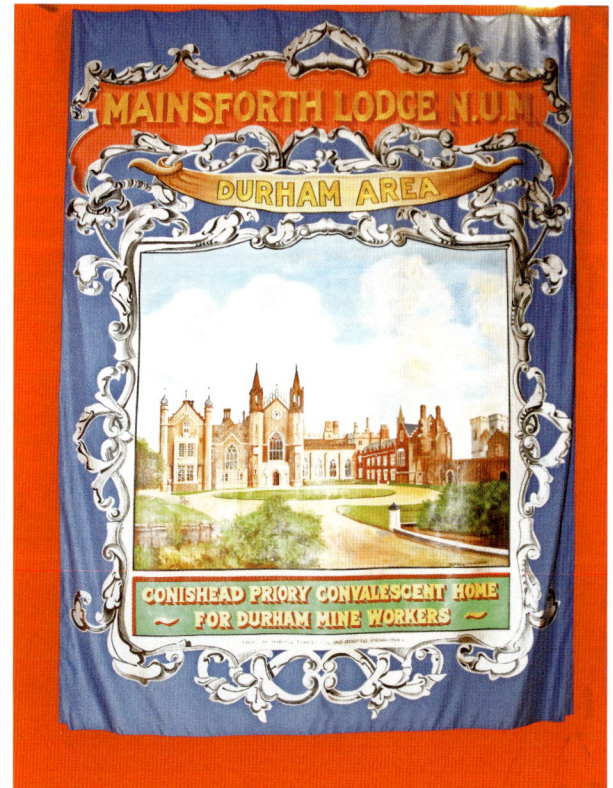

Conishead Priory:
A painting of Conishead Priory in Cumbria adorns the Mainsforth lodge banner. Conishead Priory was used as a convalescent and rehabilitation home for injured County Durham miners from 1930 to 1971. It was also used by miners suffering from various illnesses.
The recovering pitmen were able to play bowls and putting in the beautiful grounds and take trips to the Lake District. Some injured Durham men also went to The Hermitage rehabilitation centre at Chester-le-Street, opened during the Second World War, where games and other exercises were part of the therapy. The Northumberland miners' convalescent centre was at Hartford Hall, near Cramlington, where injured pitmen received similar therapy. Durham Bannermakers produced the most recent Mainsforth standard. An earlier one was made by Tutill.

His vision was our inspiration:

James Keir Hardie, first leader of the Labour Party, adorns the West Sleekburn Colliery branch banner. The date of Keir Hardie's birth, 1856, is shown below his portrait. A tribute reads: "His vision was our inspiration." West Sleekburn Colliery, near Ashington in Northumberland, closed in 1962. Keir Hardie has featured on a number of North East pitmen's banners including that of Usworth lodge.

(Photo by Derrik Scott)

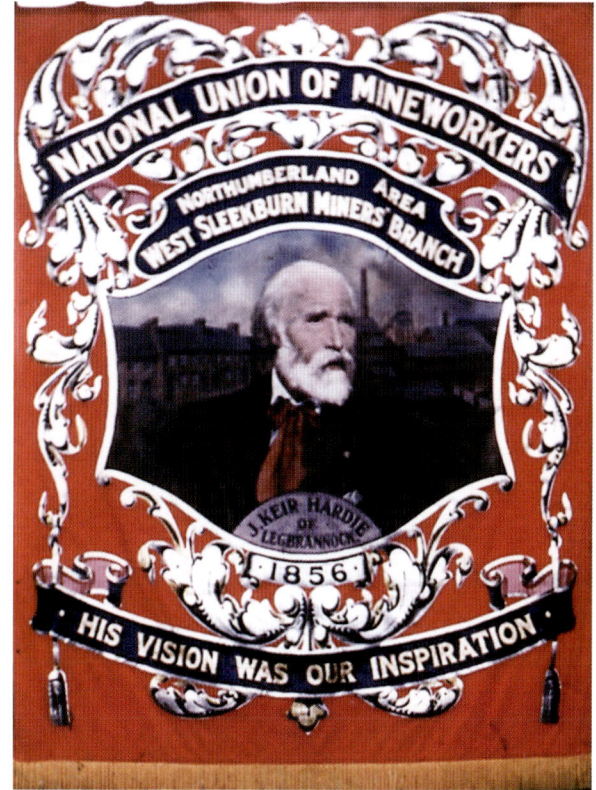

Compensation Day:

A detail from a veteran Houghton lodge banner which depicts injured miners and a doctor on 'Compensation Day'. It was made by Tutill. Houghton Colliery closed in 1981.

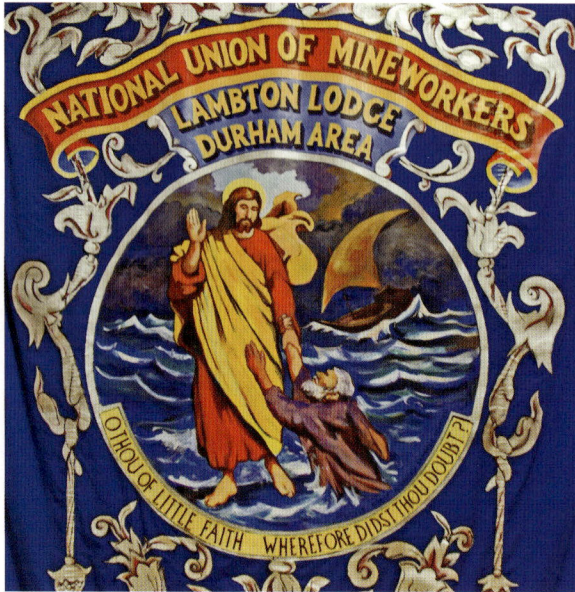

Theme from The Bible:
A Lambton Lodge banner shows Christ walking on the water. Christianity was always a strong influence in the North East pit communities. Many of the early miners' leaders and officials were Primitive Methodists. This banner was produced by Aidan Doyle, of Great Northern Banners, Newburn, and its Christian theme has featured on previous standards of the lodge.

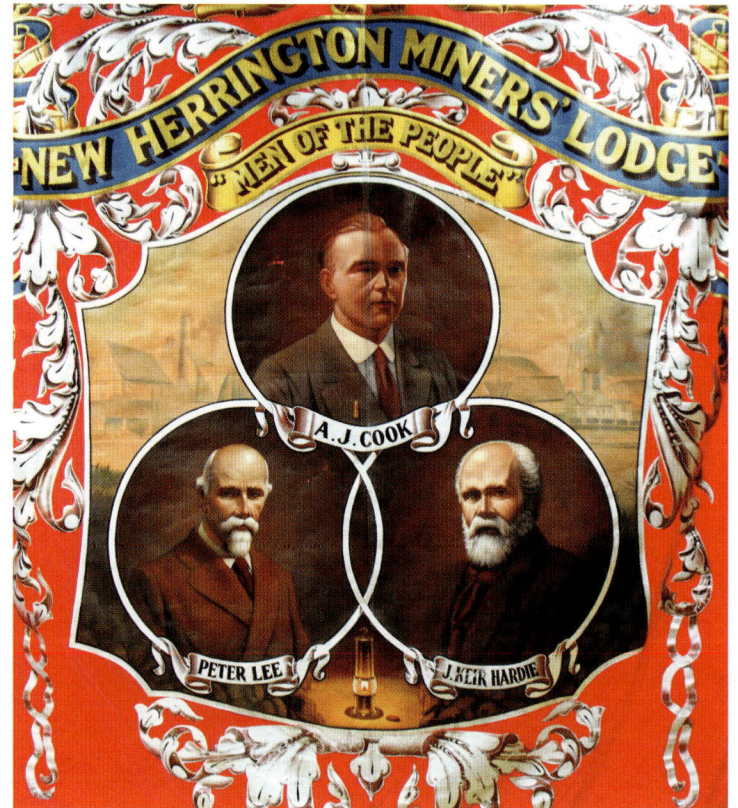

Men of the People:
'Men of the People' declares the message on the New Herrington lodge banner. Below the words are portraits of AJ Cook, who led Britain's miners during the Great Lockout and General Strike of 1926, Peter Lee, the County Durham miners' leader, and Keir Hardie, the first leader of the Labour Party.
New Herrington Colliery closed in 1985.
This banner was made by Chippenham Designs.

Peaceful message:
The Lumley 6th Pit banner carries a painting of a boy sitting on the back of a lion while a lamb looks on. This is a traditional theme, originating with Tutill ,and has been used by the lodge for many years. 'The Reign of Peace' reads the message. Lumley 6th Pit closed in 1966. This banner was produced by Aidan Doyle of Great Northern Banners, Newburn.

A memorial to miners who worked at Lumley Colliery, including those killed in accidents, can be found in the village of Great Lumley.

Among the fatalities were the 31 miners who lost their lives in an explosion of 1797 and 39 killed in an explosion in 1799.

Worldwide friendship:
Three men from across the world meet in friendship on this Dean and Chapter lodge banner. A man from Asia is shaking hands with a man from Africa, while a European man stands in the centre of the picture. "Fellowship is life, Fellowship for all" declares the message, indicating that people from all lands should unite in friendship. Dean and Chapter Colliery, at Ferryhill, closed in 1966. Norman Cornish, the renowned pitman artist, worked at Dean and Chapter for many years. The banner was produced by Durham Bannermakers.

Unveiling a banner:

The unveiling of a new banner was always a special occasion for the miners' lodges of County Durham and Northumberland. These ceremonies are still taking place today. The colourful standards are a symbol of each former pit community's pride and spirit as well as its mining heritage.

An example is the latest banner of the Usworth miners' union lodge, which was unveiled in 2015 at the Top Club, Sulgrave Village, Washington.

The banner was produced by Durham Bannermakers, a small, family firm which has created many standards for the Durham miners and other organisations. This icon of trade unionism and of the community was made by Emma Shankland and Edgar Ameti, who run the firm.

Emma, Edgar and Usworth Banner Group chairman Les Simpson unveiled the standard in front of an appreciative audience at the club.

The new banner depicts the "Three Graces" of Love, Friendship and Truth, which are represented by female figures. Accompanying the "Graces" are two miners. The design is based on a traditional motif which has been a favourite with some Durham miners' lodges over many years.

The other side of the standard shows views of the Usworth Colliery pithead, the local primary school and the community's Miners' Hall. Occupying a central position is the painting of a sculpture depicting a miner, his wife and son. The sculpture, by Carl Payne, was installed outside the bus station at Concord, Washington. The message on this side reads: 'In Family We Stand United'.

Afterwards, Dave Hopper, general secretary of the Durham Miners' Association, unveiled a plaque at the club in memory of the 42 miners who died in the Usworth Colliery disaster of 1885. The Ellington Colliery Band played the miners' anthem, Gresford, which was composed in the 1930s by Tyneside miner Robert Saint, who worked at Hebburn Colliery.

The late Dave Hopper, who served as general secretary of the Durham Miners' Association for many years, was a miner at Wearmouth Colliery. He is still remembered for his untiring efforts to improve the welfare of his fellow miners and their communities. Mr Hopper passed away in 2016.

Come let us reason together:

The Monkwearmouth lodge banner celebrates the cancellation of the Bond at the colliery as the result of a court case in 1869. The Bond was a hated annual contract which miners were forced to sign in the 18th Century and first half of the 19th Century. It bound the pitman to his employer for a year and the penalty for breaking it was a fine or imprisonment. However, the employer was not legally obligated to give the miner any work during that year.

The pitmen at Monkwearmouth (also known as Wearmouth) Colliery in Sunderland had gone on strike in a bid to win better wages after a period of severe reductions. Four of the men were arrested and charged with breaking the Bond by leaving their work. They appeared before magistrates in Sunderland.

At the hearing, W.P. Roberts, a brilliant defence lawyer who became known as the 'Pitmen's Attorney General' or the 'Pitmen's Attorney' for his dedicated work for the miners, argued that if a man was unable to read, then the mine owners should prove that the terms of the annual Bond contract had been read to him so that he fully realised the legal conditions under which he was bound.

Roberts revealed that at least one of the men could not read or write and the terms of the contract had not been read to him. He had signed the contract with his mark. The barrister contended that a man who was unable to read the Bond could not be bound by its terms. Roberts had exposed the legal weakness of the Bond system.

The prosecution was thrown into disarray and the mine owners now agreed to a proposal to cancel the Bond if the men left their colliery houses. This the four men did and moved into makeshift accommodation rather than live under the "iniquitous" contract. Other miners at the colliery now joined the revolt against the Bond and the owners realised they could not operate the mine unless they cancelled the Bond for all the pitmen.

The Monkwearmouth banner shows the courtroom scene as Roberts argues his case. Below the picture the words read: 'Come let us reason together'. The message stresses the need for negotiation as a way of resolving disputes. The Monkwearmouth case gave a tremendous impetus to the cause of trade unionism in the County Durham pits and the Durham Miners' Association was formed later that year. The leader of the Monkwearmouth strike, John Richardson, lost his job, but played an important role in setting up the association. The banner was produced by Red Wedge of Brighton.

We succour the widow and orphans:
A veteran Black Prince lodge banner carries the message: "We succour the widow and orphans." A pitman representing the union is seen comforting a widow and children at the graveside of a miner who has been killed in a pit accident. The message indicates that the union will offer support to bereaved families should their husband and father lose his life. This Tutill banner dates to the 1920s. Black Prince Colliery, near Tow Law in County Durham, dated back to the 1840s. It closed in 1933.

Three men of merit:
This Craghead Lodge banner is pictured resting against the fence at the old racecourse at the Durham Miners' Gala in 2007. Some of the band's instruments are resting in front. This banner pays tribute to Labour Prime Minister, Clement Attlee, NHS founder Aneurin Bevan and Arthur Horner, who led the NUM. Craghead Colliery closed in 1969. The banner was made by Chippenham Designs.
(Photo by Richard Smith)

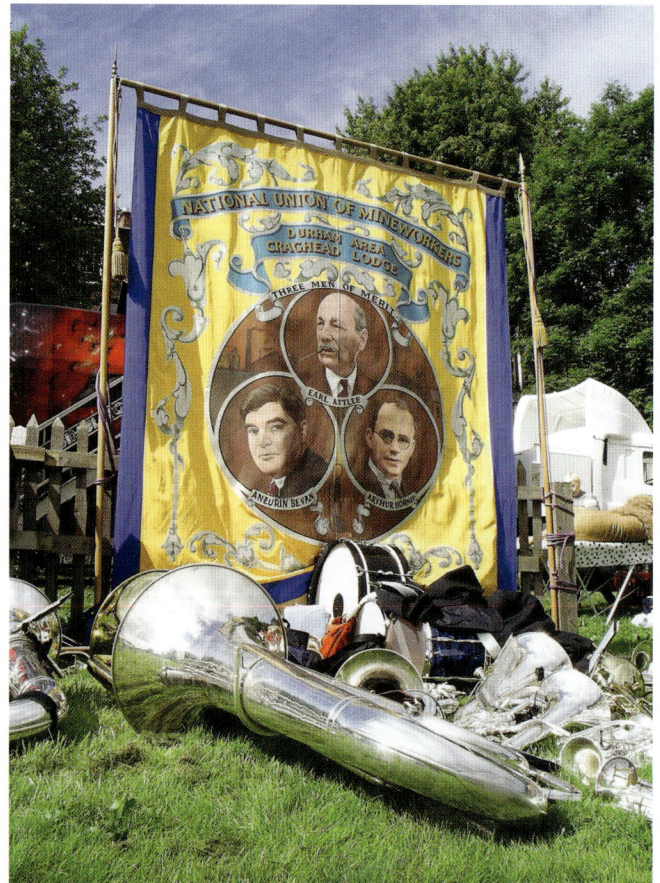

They being dead yet speaketh:

Each year, on the afternoon of the Durham Miners' Gala, the Miners' Festival Service is held in Durham Cathedral. At this special service, which dates back to 1897, new banners are dedicated and blessed by the Bishop. There is always a full congregation.

Bands and banners enter the cathedral in slow procession, the musicians playing majestic, solemn music. After the service, they march out to jaunty melodies, the congregation clapping in time to give them a rousing send-off.

Our picture shows the Haswell Lodge standard of the DMA. The banner, which dates to 1893, is displayed on a wall of the South Transept at the cathedral. The standard carries a painting of three miners' trade union pioneers. They are William Crawford, the first leader of the DMA, Tommy Ramsay, a stalwart of the Durham union during its formative years and Alexander Macdonald, a Scottish and national miners' union leader.

The motto reads: 'They being dead yet speaketh'. This modified quotation from the Bible indicates that their message of trade union principles, with its call for unity, lives on. (Photo by Derrik Scott.)

The Miners' Memorial, Durham Cathedral:

The Rev Canon Dr David Kennedy is pictured beside the Miners' Memorial in Durham Cathedral. The memorial, which takes the form of an ornate wooden fireplace, was dedicated by Bishop Alwyn Williams in 1947. It carries the inscription: "Remember before God the Durham Miners who have given their lives in the pits of this county and those who work in darkness and danger in those pits today."

Next to the memorial is a pit safety lamp, and below the lamp is the Miners' Book of Remembrance, which lists, on a colliery-by-colliery basis, numerous men and boys who died in the County Durham pits.

In Northumberland, a small Miners' Chapel of Remembrance has been created at the Church of the Holy Sepulchre in Ashington. Every year a service in memory of all Northumberland miners who lost their lives in the pits is held at the church.

The miners played a fundamental role in the history of Northumberland and County Durham. The numerous memorials, the banners, the Gala and the Picnic make it clear that they are remembered with deep respect and affection by the people of the North-East.

North-East coal mining timeline

1183 Early records of some coal mining in the *Boldon Book,* a survey of the Bishop of Durham's lands.

1239 Henry III grants Newcastle townsmen permission to dig for coal.

1260s Monks of Tynemouth believed to have mined coal and shipped it from Tyne.

1325 Document mentions ship entering Tyne with corn and returning to France with cargo of coal.

1351 Edward III grants another permission to Newcastle townsmen to dig for coal.

1605-1608 First recorded wooden waggonways in North-East, connected to River Blyth.

1708 Fatfield Colliery disaster, 69 killed, explosion.

1725-1726 Causey Arch, near Stanley, County Durham, built. Today, it is world's oldest single-span railway bridge.

1800 Phineas Crowther, of Heaton, Newcastle, invents vertical single-cylinder steam winding engine. It is adopted at many North-East collieries.

1812 Felling Colliery disaster, 92 killed, explosion.

1814 George Stephenson completes building his first steam locomotive at Killingworth Colliery workshops.

1815 Heaton Main Colliery disaster, 75 killed, inrush of water.

1815 Success Pit, Newbottle, disaster, 57 killed, explosion.

1815 Geordie and Davy safety lamps invented.

1821 First major Wallsend Colliery disaster, 52 killed, explosion.

1823 Rainton Colliery disaster, 57 killed, explosion.

1830 Jarrow Colliery disaster, 42 killed, explosion.

1831 Northumberland and Durham miners' strike under leadership of Thomas Hepburn. Employers agree to cut working hours for boys and abolish 'tommy' shops requirement.

1832 Second strike by Northumberland and Durham miners led by Hepburn. Mass evictions.

1833 Springwell Colliery disaster, 47 killed, explosion.

1835 Worst Wallsend Colliery disaster, 102 killed, explosion.

1844 Third great strike by Northumberland and Durham miners. Mass evictions.

1844 Haswell Colliery disaster, 95 killed, explosion.

1860 Burradon Colliery disaster, 76 killed, explosion.

1862 New Hartley Colliery disaster, 204 killed, shaft blocked as the result of pumping engine beam breakage and fall. Worst North-East pit disaster.

1863-1864 Northumberland miners' union, known as the Northumberland Miners' Mutual Confident Association, founded.

1864 First Northumberland Miners' Picnic. Venue: Blyth Links.

1866 First Northumberland Miners' Picnic after the election of Thomas Burt as union secretary. Venue: Polly's Folly, a field between Shankhouse and Bog Houses.

1869 Cancellation of yearly Bond at Monkwearmouth Colliery following strike and court case.

1869 Durham miners' union, the Durham Miners' Association, founded.

1871 First Durham Miners' Gala held, at Wharton Park, Durham City.

1872 Gala venue switched to racecourse by River Wear, Durham City. Every Gala since then held at this venue.

1872 Final abolition of yearly Bond.

1874 Northumberland miners' leader Thomas Burt becomes one of first two pitmen elected to Parliament.

1880 Seaham Colliery disaster, 164 killed, explosion.

1882 Trimdon Grange Colliery disaster, 74 killed, explosion.

1895 Burt Hall, Newcastle, headquarters of the Northumberland miners' union, opens.

1897 First Miners' Festival Service held in Durham Cathedral.

1898 Durham Aged Mineworkers' Homes Association founded under the leadership of Joseph Hopper.

1899 Durham Aged Mineworkers' Homes Association opens first homes at Haswell Moor.

1900 Northumberland Aged Mineworkers' Homes Association founded.

1902 Northumberland Aged Mineworkers' Homes Association opens first homes at East Chevington.

1906 Wingate Grange Colliery disaster, 25 killed, explosion.

1908 Washington Glebe Colliery disaster, 14 killed, explosion.

1909 West Stanley Colliery disaster, 168 killed, explosion.

1912 National miners' strike for minimum wage. Stoppage lasts over five weeks.

1915 Red Hill, the Miners' Hall, headquarters of the Durham miners, opens in Durham City.

1916 Woodhorn Colliery disaster, 13 killed, explosion.

1923 Medomsley Colliery disaster, 8 killed, shaft cage fall.

1924 First 'modern' pithead baths in Northumberland open at Ellington Colliery.

1925 Montagu Colliery disaster, Newcastle, 38 killed, inrush of water.

1925 Edward Pit, Wallsend, disaster, 5 killed, explosion.

1926 Great Lockout and General Strike.

1927 First pithead baths in County Durham open at Boldon Colliery.

1942 Murton Colliery disaster, 13 killed, explosion.

1947 Louisa Pit, Stanley, disaster, 22 killed, explosion.

1947 Coal mines nationalised.

1951 Easington Colliery disaster, 81 killed, explosion. Two rescue workers also lose their lives.

1951 Eppleton Colliery disaster, 9 killed, explosion.

1951 Weetslade Colliery disaster, 5 killed, explosion.

1984-1985 The Great Strike, against pit closures. Stoppage lasts a year.

1985 Last working pit ponies in County Durham retire from Sacriston Colliery.

1993 Wearmouth Colliery closes, last deep mine in Durham coalfield.

1994 Last working pit ponies in Northumberland retire from Ellington Colliery.

2005 Ellington Colliery closes, last deep mine in Northumberland coalfield and North-East.

*North-East colliery disasters occuring before 1900 with less than 40 deaths, which were numerous, are not listed in the timeline, except those occuring in the 20th Century. No accidents with under five deaths are included in the 20th Century list.

Wheels and tubs – symbols of remembrance

The full pithead winding wheel or half wheel is one of the most frequently encountered memorials to the miners and collieries of the North-East. They are to be found both north and south of the Tyne.

Examples of wheels or half wheels include those at Wheatley Hill, Thornley, Horden, West Cornforth (Thrislington) (two half wheels), Bearpark, Burnhope, South Hetton, Shotton Colliery, Sacriston, Wingate, Blackhall, Coxhoe, Fishburn (in field south of village), Wingate, Eppleton (in Hetton Lyons Country Park, Hetton-le-Hole), Seaham (with anchor on seafront), Emma (near Ryton), Chopwell, Kibblesworth, Sunderland (at Stadium of Light, site of Wearmouth Colliery), Ryhope (art work by Wilma Eaton), New Herrington (at Herrington Country Park, site of New Herrington Colliery), Murton, Ellington.

Pit wheels are also to be found at Cambois (on Northumberland coast), Dudley (at small business estate, site of colliery), Westerhope (in Newcastle, on roundabout at entrance to Westerhope Village above A1 Western bypass, in memory of North Walbottle Colliery miners), Grange Villa (next to golf course, at site of Handen Hold Colliery), Albany (Washington F Pit), Lynemouth, Newbiggin-by-the-Sea, Blyth (half wheel next to Morpeth Road Primary School and full wheel near Broadway Circle next to former miners' welfare, later council offices) and Ashington (two wheels, one in the town on Rotary Parkway and the other by the lake at Queen Elizabeth II Country Park).

The 168 miners who died in the West Stanley Colliery disaster of 1909 are commemorated by two half pit wheels sited parallel with one another in Stanley. Another wheel, at South Shields, commemorates the 51 who lost their lives in the St Hilda's Colliery disaster of 1839. At Trimdon Grange, a pit wheel stands in memory of the 74 who died in the Trimdon Grange Colliery disaster of 1882.

Another symbol of remembrance to be found in former pit communities in the North-East is the coal tub, sometimes combined with pit wheels or half wheels. Examples of tubs include those at Sacriston, Westoe (South Shields, several tubs next to roundabout), Harton (South Shields, outside Harton library), Seghill, Boldon Colliery, Sunnybrow at Willington in County Durham (Rocking Strike memorial), Bowburn, West Cornforth (Thrislington), Burradon, New Hartley (at western entrance to village), Brunswick Village (small tub near site of Dinnington Colliery), Scremerston (near Berwick, next to rugby club), Westerhope (Newcastle), Fishburn (near road to Bishop Middleham), Ellington, Choppington ('A' Pit), Nelson Village (Cramlington), Newbiggin-by-the-Sea, Bedlington (site of Doctor Pit), Spennymoor (several tubs in Jubilee Park), Bardon Mill, High Spen, Greenside, Stargate (memorial to victims of Stargate pit disaster) and West Moor.

The above list is by no means comprehensive. The authors apologise for any omissions. It is likely that more wheels and tubs will be added in future years.

Boldon Colliery:
Three coal tubs with flowers welcome visitors to Boldon Colliery, a village named after its mine. Boldon Colliery was worked from 1869 to 1982. The first pithead baths in County Durham were opened at the mine in 1927.

Fishburn:
This memorial pit wheel is an original one from Fishburn Colliery in County Durham. Situated below the wheel are inscribed memorial tablets to miners and others who worked at the colliery. They include the last pitman to leave the Fishburn mine before its closure in 1973 and a canteen lady. The wheel is situated in a recreation field a short distance to the south of the village.

Elemore:
A coal chauldron wagon stands as a memorial at the site of Elemore Colliery, now a golf course. The pit was near Easington Lane.

Sacriston:

A sculpture featuring a coal tub, safety lamp, pick and shovel at Sacriston commemorate the village colliery and its miners. This artwork is by sculptor Jim Roberts. The colliery, which was the last in County Durham to use pit ponies, closed in 1985.

Shotton Colliery:

The pit wheel memorial to all who worked at Shotton Colliery during its lifetime from 1840 to 1972. A plaque below the wheel lists the names of miners who lost their lives in accidents at the pit. The memorial is sited in Front Street, Shotton Colliery.

Westoe:

Three coal tubs commemorate Westoe Colliery and its miners. They are situated close to a roundabout near the seafront at South Shields. Fronting the tubs is a large relief sculpture of a safety lamp in the centre of a flower bed. Inscriptions on the lamp read: 'The last colliery in South Shields' with the name of the mine, and 'The past we inherit, the future we build.'

Burnhope:

A large half pit wheel at the entrance to Burhope village, near Lanchester. It commemorates the village's miners and colliery. The wheel is flanked and backed by attractive flower beds. In front is a long bench where people may sit to admire the fine view from this vantage point at an elevated road junction. Behind the wheel are two flower-filled coal tubs. Burnhope Colliery was sunk in 1850 and closed in 1949.

Wheatley Hill:

Two halves of a pit wheel fronted by flowers at Wheatley Hill. It is dedicated to all the miners of Wheatley Hill Colliery and their families. The mine opened in 1869 and closed in 1968. The memorial is opposite the site of the pit.

AUTHORS' ACKNOWLEDGEMENTS

The authors would like to thank the following for their support and kind help in the preparation of this book:

Ray Lonsdale, Bob Olley, Bryan Scott, Roy Lambeth, Kev Duncan, Michael Johnson, Jim Roberts, Michael Disley, Helen Sinden, Mark Davinson, Alan Cummings, Carl Payne, Keith Maddison, Richard Broderick, Colin Wilbourn, Cate Watkinson, John Douds, Brian Brown, Tom Maley, Ian Robertson, Jack Satterthwaite, George Robson, Dave Temple, Jack Fletcher, Dave Gibson, Les Simpson, Derek Sleightholme, Jim Wilson, Bill Bell, Olive Tindle, Barrie Ormsby, Billy Middleton, Bob Ord, Russ Coleman, Rob Walton, Pat Simmons, Phillip Blakey, Fiona Tobin, Dave Winder, John Watson, Emma Shankland, Edgar Ameti, Hugh and Lotte Shankland, Paul Pouton, the Rev Canon Dr David Kennedy, Dr Stafford Linsley, John Payton, Sue Coultard, Lynn Camsell, Norman Emery
and David Hepworth of Tyne Bridge Publishing.

The authors also acknowledge the generous help given to them by the late Derrik and Mavis Scott and the late Dr Eric Wade. There are many others who have kindly helped with this book and we apologise for any omissions.
We also extend our grateful thanks to the researchers of the Durham Mining Museum.

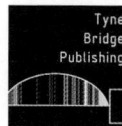

Tyne
Bridge
Publishing